MD-Vol. 88

POLYMERIC SYSTEMS

presented at
THE 1999 ASME INTERNATIONAL MECHANICAL ENGINEERING CONGRESS AND EXPOSITION
NOVEMBER 14–19, 1999
NASHVILLE, TENNESSEE

sponsored by
THE MATERIALS DIVISION, ASME

edited by
A. SAIGAL
TUFTS UNIVERSITY

K. RAMANI
PURDUE UNIVERSITY

THE AMERICAN SOCIETY OF MECHANICAL ENGINEERS
Three Park Avenue / New York, N.Y. 10016

FOREWORD

This bound volume contains papers presented at the Symposium on Polymeric Systems, organized by the Polymers Committee of the ASME Materials Division, at the 1999 International Mechanical Engineering Congress & Exposition held in Nashville, Tennessee. This symposium promotes a systems approach to processing, design and mechanics of polymeric systems, which include composites, sandwich constructs and bonded joints. While materials continue to evolve at a pace never seen before, the design, processing and mechanics of these systems are beginning to be understood more closely. A key aspect is the material microstructures are being designed simultaneously along with the processes. With this symposium we bring leaders in diverse areas of polymeric systems together to promote interdisciplinary thinking.

The papers presented in this Symposium are in the general areas of testing of polymer composites, bonded structures and composite design. Specific papers deal with mechanical properties, testing, and durability of composites, impact behavior and characterization of polymers, bonding of polymers to metals, and analysis of composite pressure vessel domes and polymer/reinforced concrete beams.

We are very pleased to see the interest in these subjects and look forward to sharing future work in these areas. Finally, we wish to extend our sincere thanks to the authors for preparing the papers on time and Ms. Barbara Signorelli for publishing this bound volume.

Anil Saigal
Tufts University
Medford, Massachusetts

Karthik Ramani
Purdue University
West Lafayette, Indiana

CONTENTS

AN ALTERNATIVE MECHANICAL PEEL TESTING METHOD
FOR COMPOSITE FIBERGLASS FOAM CORE SANDWICHES

John L. Zych, Scott A. Adamick, Daniel K. Jones, Ph.D., P.E., Dru M. Wilson, Ph.D., Wayne A. Boyer

Central Michigan University

ABSTRACT

Nondestructive and destructive testing are vital in the composite material industry to determine the mechanical properties of different designs. This paper presents a novel approach to measuring the mechanical peel strength that may be used for a variety of fiberglass foam core sandwiches. The objective of this study was to develop a testing method that was simple and easily reproduced in other laboratories. The goal was to measure the mechanical peel strength using a device that may be used for a variety of sample sizes and materials. An additional goal was to provide benchmark data on the mechanical peel strength of two particular composite designs. The new testing method may have applications to a wide range of composite structures, and it may eventually lead to a new ASTM standard.

INTRODUCTION

The American Society for Testing and Materials (ASTM) is an independent, nonprofit organization that establishes standards and specifications on methods of testing many properties for a variety of materials. The current ASTM methods for mechanical peel testing of composite materials[1] are not specifically designed for sandwich panels consisting of fiberglass skins with foam cores.

According to Shields[3], "extension of a flexible member during peeling leads to inaccurate results; the stretch forces are included in the test readings where the line of contact (at which bond failure occurs) becomes reduced. Nevertheless, the imperfect peeling action may closely simulate an actual assembly under stress, and provide a more meaningful test result." As noted by Gould[2], the accuracy of the test is a function of the strain rate and the peel angle. A new testing method is needed to address these concerns.

Testing methods are required to determine the quality and to assess the integrity of new composite designs. A new testing method may lead to improvements in the reliability and validity of testing the samples. Results may be used for improvements in manufacturing processes and quality assurance in production operations.

The new method will also provide benchmark data that will be useful for product applications. Testing of new structures is crucial in the composite material industry to determine physical properties. Problems persist because mechanical peel tests do not exist for composite materials.

METHODS

The composite samples were made of two plies of fiberglass for each face sheet, bonded to a foam core using epoxy resin.[4,5,6] The specimens were two inches wide by ten inches long, adhered to a plywood base to facilitate mounting onto the test apparatus. One inch of fiberglass extended beyond the core at the top to facilitate gripping of the upper sheet for peeling.

A prototype peeling apparatus was designed and constructed to peel samples at various angles while adding weight at a constant rate. A conceptual drawing of the device is shown in Figure 1. The mounting base was welded to a three-inch steel sheet to a one-inch angle beam at a ninety-degree angle. This vertical beam supported a pulley that could be secured at different heights in order to adjust the peel angle.

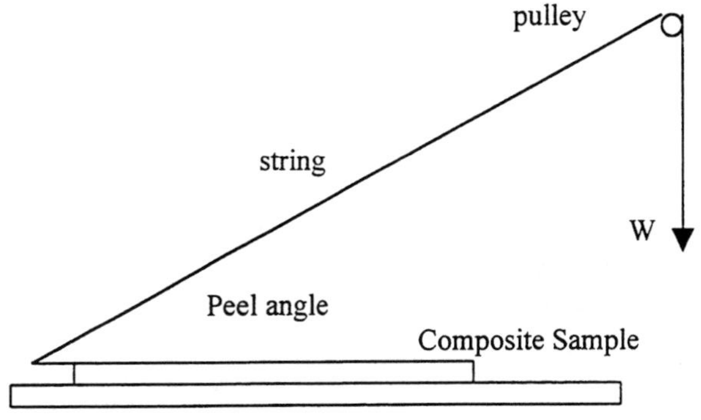

Figure 1. Peeling Apparatus

The composite samples were secured to the horizontal base while a clamp and string was attached to the upper edge of the fiberglass. The string was inclined at two angles to determine which was the most effective at peeling the fiberglass off of the foam. This string was guided around a pulley and attached to the handle of a bucket. Water was added at a constant flow rate into the bucket, increasing the weight until the adhesive failed. At failure, the weight applied and the angle with which it was tested were recorded. Twelve samples at two angles were tested. The intent was to provide benchmark data and to determine if the mechanism was reliable and repeatable.

The following steps were taken to test the samples. Foam core sandwich specimens were placed into the peel testing mechanism. With the protractor, the peel angle was set to the desired value (45° or 60°.) A bucket of water, attached to the far end of the string, was used to add weight to the sample. Water was added at a constant rate of one liter per minute via siphoning water from a reservoir. As soon as the flow began, a stopwatch was started to time the flow. The weight of water was added slowly, and failure occurred rapidly, when the fiberglass pulled away from the foam core. The time at failure was recorded, and the corresponding weight of water in the bucket represented the failure force.

RESULTS

Peel force was measured for a total of 24 samples. Twelve of the samples were grey in color, and the other twelve were natural in color. The core of the grey samples was polyvinyl chloride (PVC) foam with a density of 80 kg/m^3, and the core of the natural samples was PVC foam with a density of 100 kg/m^3.

Table 1. Measurements of Peel Force

Composite Type	Peel Angle (degrees)	Peel Force (Average±SD) (N)
Grey	45	79.5 ± 11.1
White	45	106.1 ± 36.6
Grey	60	95.0 ± 10.4
White	60	119.0 ± 9.8

Six of each sample type were tested at a peel angle of 45°, and six of each were tested at 60°. For each composite type, at each angle, the peel force is listed in Table 1. Each row represents a group of six samples.

INTERPRETATIONS

At each peel angle, the grey composite had lower peel forces than the natural samples. This may be attributed to a weaker core sample or a weaker bond between the fiberglass and the foam core.

For both composite types, the higher peel forces resulted at higher peel angles. This was unexpected, since the higher peel angles had larger vertical force components, which are intuitively associated with peeling. This suggests that the horizontal force component of the peeling force plays an important role in peeling.

The variance among samples was due to many variables that are included when making the samples. For example the samples were made by hand, therefore an uneven amount of resin could have been applied or else damaged while cutting the parts to size. Relatively high standard deviations in the peel force measurements may be attributed to the small number of samples and the inherent variance among the composite constructions.

CONCLUSIONS

To improve the technique, several steps could have been performed to achieve a more consistent reading. The clamp that attached to the fiberglass was observed to bend on the sides which caused an uneven peel. To fix this, a more rigid clamp could be used. Peeling the sandwich evenly along the section may prevent one corner from peeling before the other.

An exact flow rate for each sample would have also improved our times. The level in the reservoir varied slightly from sample to sample, which would cause the data to not be exact.

In conclusion, the apparatus was effective at peeling every sample without breaking the fiberglass. The preliminary results suggest that, although improvements are possible, the new peeling force apparatus provided useful pilot data for measuring the peel force at different angles.

ACKNOWLEDGMENTS

The authors would like to thank the Industrial and Engineering Technology Department at Central Michigan University for providing raw materials and laboratory equipment. The authors also appreciate the instruction and support provided by Bill Dekryger, David Spitler, Duane Malburg, and John Nee.

REFERENCES

[1] Standard Test Method for Climbing Drum Peel for Adhesives, American Society for Testing and Materials, ASTM Designation: D 1781-93, pp. 93-96.

[2] Gould F., Robert <u>Contact Angle Wettability and Adhesion</u>. Advances in Chemistry Series, American Chemical Society, 1964.

[3] Shields, J., <u>Adhesives Handbook</u> International Scientific Series #55, CRC Press, 1970

[4] Rosato, D.V., <u>Designing with Reinforced Compsites: Technology-Performance-Economics</u>, Hanser Publishers, New York, 1997.

[5] Colling, D.A., and Vasilos, T., <u>Industrial Materials: Polymers, Ceramics, and Composites</u>, Vol. 2, Prentice Hall, Inc, Englewood Cliffs, New Jersey, 1995.

[6] Foreman, C., Advanced Composites, IAP, Inc., Casper, WY, 1990.

EXAMINING MULTIAXIAL IMPACT BEHAVIOR OF POLYMER MATERIALS

G. E. Lawrence, A. Saigal, M. A. Zimmerman*, R. Greif and Y. Duan

Dept. of Mechanical Engineering
Tufts University
Medford, MA, USA 02155

*Lucent Technologies
North Andover, MA, USA 01845

ABSTRACT

The analysis of multiaxial impact of polymer disks is considered. The calculation of impact displacements and stresses is provided. Finite element simulation results are compared to experimental data. The results from simulations of various impact velocity and mass are given for a constant disk thickness. Results from simulations of various disk thickness for constant impact mass and velocity are shown as well. The plasticity failure model used in FEA simulation of impact is quantified for the application with the tested polymers. It is shown that strain rate material dependence is an important factor in accurately modeling impact response of polymers.

NOMENCLATURE

D - Plate Constant
E - Modulus of Elasticity
M - Bending Moment
S - Stress
W - Load
Y - Deformation
a - Disk Radius
f - Isotropic Yield
h - Drop Height

r - Contact Radius
t - Disk Thickness
$\bar{\varepsilon}^{pl}$ - Plastic Strain
$\dot{\varepsilon}^{pl}$ - Plastic Strain Rate
f^{α} - Definable Field Variables
σ - Uniaxial Stress
θ - Temperature
ν - Poisson's Ratio

INTRODUCTION

Bogdonavich and Iarve (1992) refer to the lack of development in the area of impact deformation and failure theory for plates. As such, experimentation and simulation are commonly employed to understand impact behavior. ISO 6603/2 (1989) is a test procedure for multiaxial impact testing of polymeric materials. It is often employed to examine the impact behavior of a polymer material. Specifically, it is employed to determine the amount of impact energy absorbed at failure.

The destructive nature of the ISO test combined with the cost of producing test specimens indicates the potential usefulness of FEA simulation. Development costs of a product employing a polymer can be reduced through optimized simulation and testing. However, the highly nonlinear nature of impact and the development of precise material and failure models for FEA are both obstacles to accurate simulation.

The accuracy of simulations is examined below with consideration of different impact failure models. The impact failure model suggested by General Electric (1998) defines failure by excessive deformation and stress. In addition, some potentially applicable methods within the commercial FEA software used for simulations are examined, such as rate dependent hardening, and the accuracy of the method chosen is revealed.

PROBLEM DESCRIPTION

When considering polymer materials for an electronics enclosure application, several materials may be considered depending on different application variables. Recently, an electronics manufacturer developed an application requiring out-door weatherability, impact resistance, and radio-frequency noise resistance. Based on these needs, several polymers were considered and tested. Commercial grades of Polybutylene Terephalate (PBT), Acrylic-Styrene-Acrylonitrile (ASA), and Acrylonitrile-Butadine-Styrene (ABS) were targeted for impact testing. The tensile mechanical properties collected using low strain rate tensile tests are listed below in Table 1. Ten samples of each material were subjected to impact under the guidelines of International Standards Organization document ISO6603/2.

Table 1- Tensile properties of polymers investigated

Material	Modulus (MPa)	Yield Strength (MPa)	Break Strength (MPa)	Elongation at failure (%)
PBT1	2428	56.9	45.2	109.6
ASA1	2079	41.2	30.7	16.6
ABS1	2339	43.1	38.8	6.2
ABS2	2523	47.2	35.6	27.3
ABS3	2243	42.2	31.9	25.9

FINITE ELEMENT MODEL

To better understand the impact behavior of the polymers, finite element models were used to simulate the impact test ISO 6603/2. ABAQUS/Explicit, a commercial explicit finite element software, was used to create the model and simulate the impact. The disk impact was modeled using quarter symmetry using 3-dimensional shell elements for the disk and 3-dimensional rigid elements for the impact sphere. Boundary conditions were applied to the disk to simulate the clamped configuration of ISO6603/2, and the impacting sphere was given an initial velocity condition. The disc diameter is 40 mm. The projectile hemisphere has a radius of 10 mm and a mass of 5 Kg. The height of the drop is chosen to provide adequate kinetic energy to cause failure. Figure 1 shows the undeformed model.

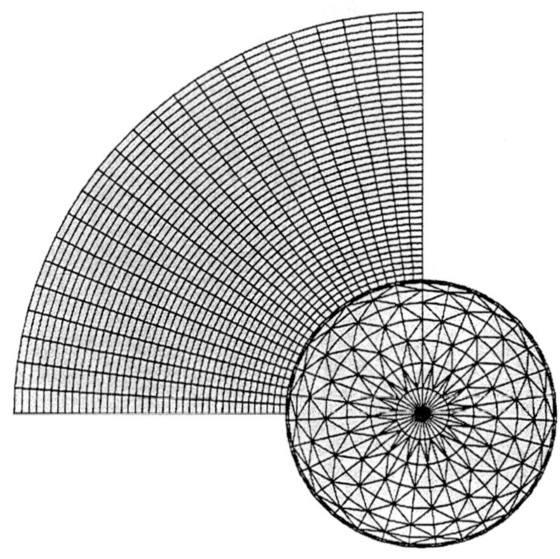

Fig. 1 – Undeformed Model of ISO6603/2 Geometry

The FEA code requires selection of a failure model. Due to the apparent ductility of the materials tested, the code's metal plasticity failure model was chosen. This model requires tensile modulus, yield strength, and elongation and stress at failure. The model creates a damage factor, which is increased as the equivalent plastic strain increases. When the equivalent plastic strain equals the prescribed strain at failure, the model eliminates that material point. The damage parameter, ω, is defined as

$$\omega = \sum \left(\frac{\Delta\varepsilon^{pl}}{\varepsilon_f^{\,pl}} \right)$$

where,

$\Delta\varepsilon^{pl}$ is an increment of the equivalent plastic strain;

$\varepsilon_f^{\,pl}$ is the strain at failure.

The failure of the disk is depicted in Figure 2 below. The impact sphere completely punctures the disk as expected at the impact energy of 250 J. However, it is the energy consumed by the disk at failure, which is the plastic deformation energy calculated by the FEM model, that is of importance. Table 2 shows the experimental and simulated results for each material. Note that the experimental and simulation failure energies are not in good agreement. The best simulation result is less than 50% of the experimental value. Except for ABS1, none of the simulation results would be statistically significant because all are nearly three standard deviations away from the experimental values.

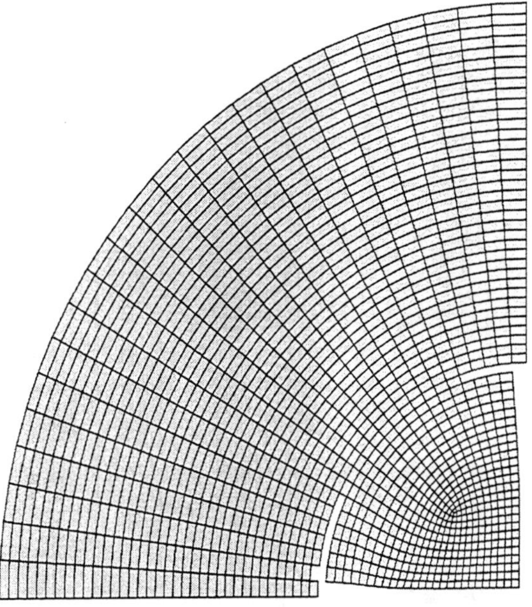

Fig. 2 - Failure of Modeled Disk

The average failure energy and standard deviation were noted for each material. Failure energies ranged between 2 and 33 J. The tensile mechanical properties listed in Table 1, collected using low strain rate tensile tests, do not obviously explain the wide variation in failure energies. The greatest resistance, seen in an impact modified PBT listed as PBT1, is expected due to the great ductility indicated by a 109% elongation at failure. ABS1, a recycled ABS, displayed the lowest impact failure energy and has the least ductility with an elongation of 6.2%. ABS2, virgin ABS of the same grade as ABS1, and ABS3, a reinforced ABS, performed far better than the recycled material. This despite the fact that all three materials have nearly the same elastic modulus and tensile strength. It is possible that recycling affects some aspect of the material, which causes reduced impact performance. However, it should be noted that while the elongation at failure of ABS2 and ABS3 are over 4 times that of ABS1, their impact resistance is over 6 times greater. If the energy values were not equal to each other, but the offset percentage was constant, a scale issue could be detected. The lack of such a scale or constant offset indicates a need for deeper examination of the issue.

Table 2- Failure energies for each material tested

Material	Experimental Failure Energy (J)	Experimental Standard Deviation	Simulation Failure Energy (J)
PBT1	33.1	0.7	16.0
ASA1	18.5	4.8	2.3
ABS1	2.2	0.6	1.3
ABS2	12.6	2.8	3.6
ABS3	14.7	4.7	3.4

A further analysis was done to examine the variation in failure energy as a function of disk thickness. The procedures listed in ISO 6603/2 provide ranges for several test parameters. For the data provided, disk thickness was 2.0 mm. As seen in Table 3 below, as the thickness of the PBT disk increases, the failure energy increases non-linearly. It appears that only an increase in simulated thickness approaching the maximum specification of the ISO test (4.0 mm) would result in the simulation and experimental failure energies to be approximately equal. As such, theoretical models must be examined to understand the lack of agreement of the simulations with experimental data.

Table 3- Failure energy for varied disk thickness

Material	Disk Thickness (mm)	Simulation Failure Energy (J)
PBT1	1.0	5.6
PBT1	1.5	10.0
PBT1	2.0	16.0
PBT1	2.5	18.4
PBT1	3.0	24.8

A THEORETICAL MODEL

General Electric plastics division provides a theoretical solution for impact of polymer materials. In an example of impact resistance of Lexan® (GE Plastics, 1998), formulas from Roark and Young (1996) are used to calculate the dynamic stress on a rectangular skylight due to impact by a mass dropped from a height. The stress value is compared to the yield strength of the material to provide a failure criteria. This procedure is modified using Roark and Young's circular plate formulas. The equations below govern the model, where the plate constant, D, depends on modulus, thickness and Poisson's ratio as

$$D = \frac{Et^3}{12(1-v^2)} \qquad (1)$$

static deformation at the center of the disk, Y_s, depends on load weight, disk geometry and plate constant as

$$Y_s = \frac{-Wa^2}{16\pi D} \qquad (2)$$

bending moment, M, depends on load weight, Poisson's ratio and the disk radius over contact radius as:

$$M = \frac{W}{4\pi}(1+v)\ln\frac{a}{r_0} \qquad (3)$$

static stress, S_s, depends on bending moment and thickness as

$$S_s = \frac{6M}{t^2} \qquad (4)$$

The dynamic stress, S_i, depends on static stress, static displacement and drop height as

$$S_i = S_s(1+\sqrt{1+\frac{2h}{Y_s}}) \qquad (5)$$

and the dynamic displacement, Y_i, depends on static displacement and drop height as:

$$Y_i = Y_s(1+\sqrt{1+\frac{2h}{Y_s}}) \qquad (6)$$

To quantify the performance of the simulations, the failure energy results were substituted into the equations above. The failure energy output was used to provide a velocity of impact for the given mass. As the formulas use drop height, the velocity was converted to a height using free fall dynamics. Table 4 shows the predicted stress (S_i) and the yield stress for disks of each material, having the same geometry, using the adapted GE criteria.

The GE method is also not in good agreement with the experimental data. Examination shows that the GE method is an approximation, which is best used to examine if under the give conditions, impact will result in plate failure.

Table 4- Theoretical dynamic stress and yield stress

Material	Minimum S_i (MPa)	Yield Stress (MPa)
PBT1	263.6	56.9
ASA1	95.7	41.2
ABS1	77.4	43.1
ABS2	130.0	47.2
ABS3	119.6	42.2

The GE model is based largely on elastic modulus (E), and the metal plasticity model of the solution uses data from the constant, low strain rate tests. Therefore, in cases of high strain impact, specifically cases where the response to impact depends on strain rate, the metal plasticity model of the FEA code and the GE method are not appropriate and must be modified.

ALTERNATIVE MODELS

When describing a strain rate dependent yield model, the user's manual for the FEA code by HKS, Inc. (1997), states " Many materials show an increase in their yield strength as strain rates increase; this effect becomes important in many metals and polymers when rates range between 0.1 and 1 per second, and it can be very important for strain rates ranging between 10 and 100 per second, which are characteristic of high-energy dynamic events or manufacturing processes." The strain rates expected in the 2 mm thick disks, subjected to 250 J impact energy, requires further examination of strain rate dependency.

The dependency model, based on isotropic hardening, allows the plastic strain to surpass the incipient yield point. The yield function for isotropic hardening is shown below as Eq. 7. The function is a symmetric second order tensor based on the equivalent yield stress σ_i, which can be defined for uniaxial tension, compression, or, shear yield stress. The stresses are functions of plastic strain, plastic strain rate, temperature, and other field variables.

$$f(\sigma) = \sigma_i(\overline{\varepsilon}^{pl}, \dot{\varepsilon}^{pl}, \theta, f^\alpha) \qquad (7)$$

Implementation of this model requires the user to input stress-strain tables for several strain rates and temperatures. The extensive amount of data needed to simulate materials with these properties led to an additional investigation. The impact velocity was varied for one material of constant thickness in order to investigate the effect of input strain rates. As seen below from the data in Table 5 for strain-rate independent materials, the failure energy is constant for disks impacted at different velocities. This confirms that the simulation model can not respond in a manner consistent with the materials in the experiments.

Examination of the rate dependent properties of a grade of LEXAN 101 polycarbonate from GE, highlights the manner in which material rate dependency may influence impact behavior (Figure 3). The curves are tensile stress vs. strain for various strain rates and temperatures. Universally, the elongation at failure is higher for samples tested at higher strain rates. In fact, the plastic strain is over 60% greater for those cases. This increased plastic strain would increase the energy required to break the material, which means a direct increase in impact resistance. Typical strain rates encountered in the impact simulations in our investigation were of the order of 10^2 /s.

CONCLUSIONS

The experimental data gathered using ISO6603/2 to test the polymers indicate rate dependent hardening. Clearly, the conclusion is that both the theoretical and simulated models applied are not appropriate to accurately quantify the rate dependent impact response of these polymers. In the future, further experimentation will be required to provide stress-strain curves for several strain rates similar to those for Lexan in Fig. 3. Once this additional data is gathered, a more accurate simulation can be achieved through the rate dependent hardening model.

REFERENCES

Bogdonavich, A. E.,and Iarve, E. V., 1992, "Numerical Analysis of Impact Deformation and Failure in Composite Plates," Journal of Composite Materials, Vol. 26, No. 4, pp. 520-545 .

General Electric Plastics, 1998, "Example of Impact Calculations," General Electric Co., Pittsfield, MA, www.ge.com/plastics/lexan/design.

Hibbitt, Karlsson & Sorensen, Inc., 1997, "Rate Dependent Yield," ABAQUS/Explicit User's Manual, Vol. 1, Version 5.7, pp 10.2.3-1-10.2.3-3.

International Organization for Standardization, 1989, "Plastics- Determination of Multiaxial Impact Behaviour of Rigid Plastics - Part 2: Instrumented Puncture Test," International Standard ISO6603-2,First Edition, ISO Geneva, Switzerland, pp. 1-12.

Young, W. C., 1996, "Roark's Formulas for Stress & Strain," McGraw-Hill Publishing Corp., New York, NY, Sixth Edition, pp. 391-440.

Table 5 – Simulated failure energy for various rates

Material	Impact Velocity	Failure Energy (J)
ASA1	5 m/s	2.3
ASA1	10m/s	2.3
ASA1	15 m/s	2.3

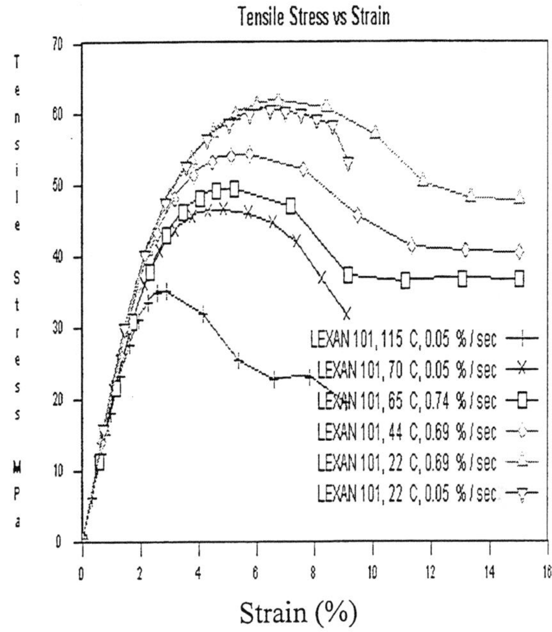

Fig. 3 - Stress-strain behavior of GE Lexan at different temperatures and strain rates

EFFECT OF MATERIAL PROPERTIES ON STRESS INTENSITY FACTOR MEASUREMENT BY THE FROZEN STRESS METHOD

C. W. Smith
Dept. of Engineering Science and Mechanics
Va. Poly. Inst. and State University
Blacksburg, VA 24061

C. T. Liu
Air Force Research Laboratory
10 E. Saturn Blvd.
Edwards AFB, CA 93524-7680

Abstract

After briefly outlining methods of analysis, this paper summarizes a study of the influence of material properties on the stress intensity factor and it's distribution along the borders of cracks within and parallel to bondlines for both through and part through cracks in plates. Examples of test results are included to support summary assessments.

Introduction

Historically, when photoelasticians conduct frozen stress tests on complex shapes requiring bonded joints, it has been considered good practice to avoid taking data along these joints due to residual stress fields resulting from mismatched properties between the photoelastic material and the adhesive. In recent years a variety of adhesives have been developed which significantly reduces this problem when bonding two pieces of the same material but for bimaterial joints, problems still exist.

It is not unusual then, that cracks often develop within or parallel and near to the bondlines. Beginning in 1993, the authors undertook a study involving the determination of the stress intensity factor (SIF) distribution along the border of a through crack in the bondline of a bimaterial specimen. Since it was virtually impossible to obtain optical quality stress freezing materials with the same critical temperature (T_c) but varying modulus, a three specimen test method was developed (Smith, Finlayson and Liu, 1997) for separating residual stress from modulus mismatch and T_c mismatch and the modulus of a common stress freezing material (araldite) was altered (without changing it's T_c) by seeding it with aluminum powder. By bonding the seeded araldite to standard araldite, a model was made which contained no T_c effect and it was used to measure the T_c effect in cracked bimaterial specimens containing materials with a T_c mismatch.

Stress Freezing and Bimaterials

The stress freezing process involves heating the transparent photoelastic material (an epoxy) which, in simplest context, behaves in a Kelvin like manner at room temperature (Fig. 1) to a critical temperature (T_c) which is near it's glass transition temperature. When then loaded above T_c, the viscous coefficient (μ) vanishes and the material behaves linearly elastic but with a modulus of only about $1/600^{th}$ of its room temperature (R.T.) value but is some 20 times more sensitive to the photoelastic effect. By then cooling slowly under load back to R.T., the photoelastic stress

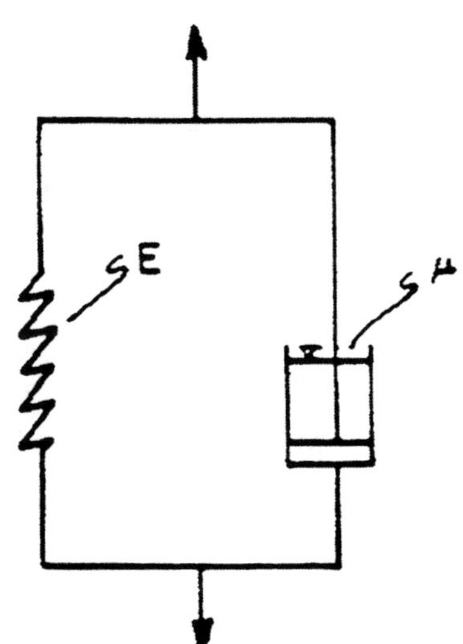

Figure 1 Kelvin Model

fringes induced by loading above T_c are retained in the model. When unloaded, the recovery is negligible due to the higher modulus and lower stress fringe sensitivity at R.T. The model may then be sliced into layers without disturbing this fringe pattern for analysis.

When the cooling cycle is applied to a stress freezing material, its modulus and thermal expansion coefficient changes approximately as shown in Fig. 2, where $T_c \approx 275°F$.

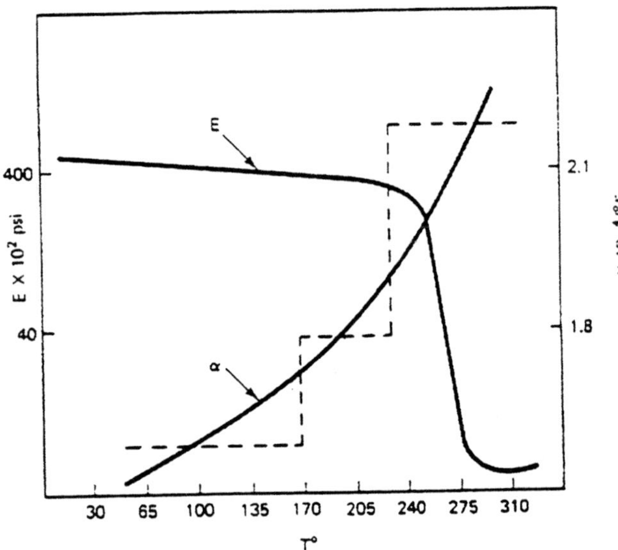

Figure 2 Effect of T_c on Material Properties

When a bimaterial specimen consisting of two stress freezing materials with differing T_c values is thermally cycled as above, a complex interaction results. It is the objective of this study to measure the effects of these interactions. In order to achieve this, two model material combinations were employed.

Bimaterial Models B consisted of two, off the shelf stress freezing photoelastic materials, the properties of which are given in Table I and show a 25°F T_c mismatch. Most of the tests on cracks parallel to and within the bondline were conducted with Models B.

Experimental Procedure

The three specimen test method described above consists of 1) a homogeneous edge cracked specimen, 2) a homogeneous bonded edge cracked specimen and 3) a bimaterial edge cracked specimen. The cracks were made with a Buehler saw and were aligned to

Table I Material Properties
Models B

Models B	T_c	E_{Hot}	f
PLM-4B	180°F	(12.96 Mpa)	(420.3 Pa-m)
PSM-9	205°F	(47.67 Mpa)	(520.1 Pa-m)

These materials have the same thermal coefficients at room temperature (T_0) and critical temperature (T_c). They were $\alpha_{\text{RT}} = 39 \times 10^{-6}/°F$ and $\alpha_{\text{CT}} = 90 \times 10^{-6}/°F$ respectively. Bonding agent was liquid PSM-9. f = material fringe value.

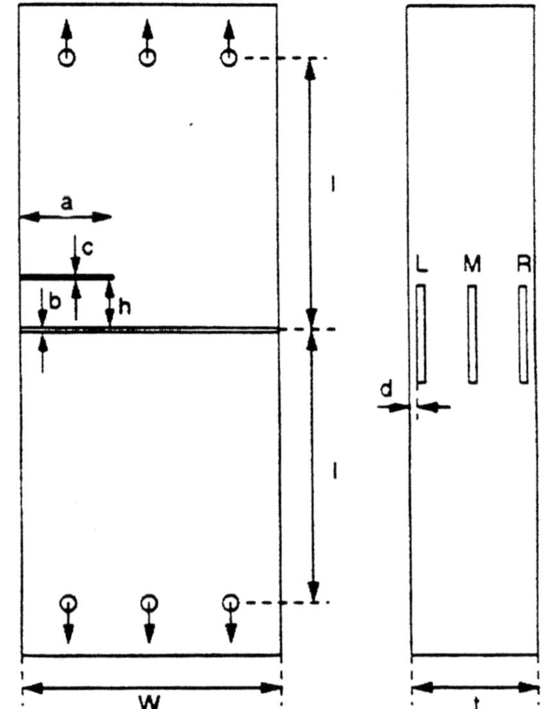

h = 0, 0.125, 0.250, 0.500 in. (0, 3.18, 6.35, 12.7 mm.)

a = 0.375 in. (9.53 mm.)   b = 0.012 in. (0.305 mm.)
c = 0.012 in. (0.305 mm.)   d = 0.036 in. (0.914 mm.)
s = 0.036 in. (0.914 mm.)   t = 0.500 in. (12.7 mm.)
l = 4.50 in. (114.3 mm.)   W = 1.50 in. (38.1 mm.)

Figure 3 Test Specimen Dimensions

the bondline. For cracks in the bondline itself, they filled the bondline area so that only test data were taken in the adherends themselves. The test specimen dimensions are given in Fig. 3 for the through thickness cracks and in Fig. 4 from the part through

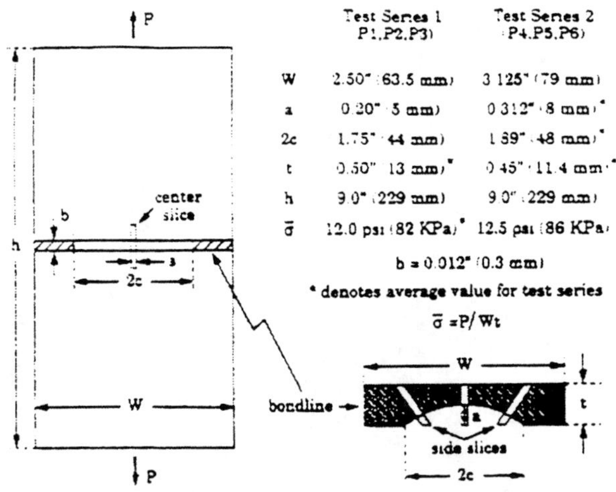

Figure 4 Test Specimen Dimensions and
Slice Locations

cracks. Stress freezing tests were conducted on Models B using all three specimen types for the values of h given in Fig. 3 and each specimen was sliced mutually orthogonal to the crack border and the crack surface at intervals and each slice was analyzed photoelastically. Slice orientations for part through cracks are shown in Fig. 4 along with test specimen dimensions for test series 1 and 2. By combining Post partial mirror fringe multiplication with the Tardy method, fringe orders were read to $1/50^{th}$ of a fringe.

<u>Algorithms</u>

Once photoelastic data are obtained, an algorithm is required to convert the optical data into stress intensity factors (SIF's). For thick incompressible materials, the interface fracture equations (Hutchinson and Suo, 1992) reduce to the classical form for homogeneous materials, allowing K_1 and K_2 separation. Such conditions existed in the current experiments and the K_1 algorithm used is developed in the Appendix. The K_2 algorithm is found in Smith and Kobayashi, (1987)

<u>Results</u>

The through thickness crack test results from Models B revealed little effect of the T_c mismatch, yielding Mode I values which agreed with the homogeneous test result for $h = 12.7$ mm and 6.35 mm. However a significant increase of 30% resulted in K_1 for $h = 3.18$ mm and a measurable amount of K_2 was

observed (shown by the rotation of the figure eight near tip fringe loops in Fig. 5).

In order to determine whether these effects were due to T_c mismatch or modulus mismatch, a new set of models were prepared by combining araldite adherends with araldite adherends seeded with aluminum particles to increase the modulus without altering T_c. Called Models A, their properties are given in Table II.

Table II Material Properties
Models A

Models B	T_c	E_{Hot}	f
Araldite	240°F	(18.60 Mpa)	(286.9 Pa-m)
Aral-Alar	240°F	(36.88 Mpa)	

Matched thermal coefficients at 68°F.; $\alpha = 15.3 \times 10^4$ per °F. At critical temperature (T_c) thermal coefficients were $119 \times 10^{-6} \pm 20 \times 10^{-4}$ per °F. f = material fringe value.

Using Models A, the test geometry with $h = 3.18mm$ was replicated and tested as before. The resulting K_1 was not elevated and in fact was identical to the homogeneous K_1 value with no K_2 value present (see Fig. 6). This suggested that the K_1 elevation for $h = 3.18$ mm was due to T_c mismatch, not modulus mismatch. A typical result showing the SIF distribution through the thickness and confirming the lack of SIF elevation due to modulus mismatch is shown in Fig. 7. In all of the studies, the SIF values were normalized with respect to a theoretical 2D estimate as shown in Fig. 7 to account for any effects due to the use of an artificial crack instead of a real crack. Moreover, the experimental K_1 values are the average from slices L, M, R. A final test was conducted with Model B composition with $h = 0$. The result is shown in Fig. 8. There is no local fringe rotation indicating no K_2 present and the K_1 values obtained on either side of the crack are within experimental scatter of 5%. It should be noted that the fringe pattern is not symmetrical with respect to the crack plane because the material fringe values of the two adherends were not the same (Table I). This suggests that Model B can be used to evaluate K_1 distributions in through cracks.

For the part through cracks, there was little variation between the side slices and center slice results

11

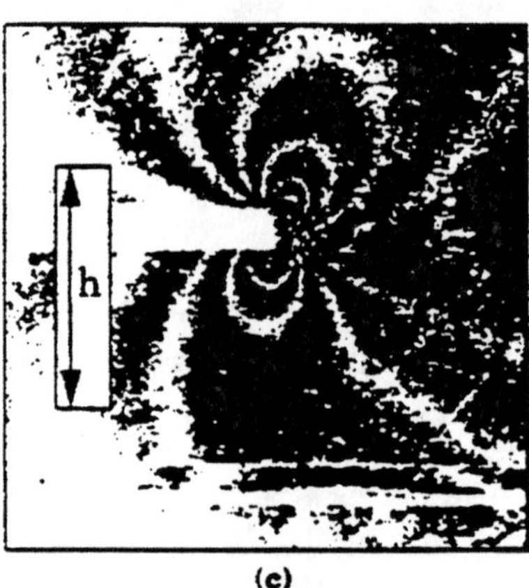

a) Pre-load fringe field. (tt)
b) Post-load fringe field. (tt)
c) Middle slice fringe field.
Magnification factor = 8.5
tt = through-thickness
dz = data zone
h = 0.125 in. (3.18 mm)
θ_R = Reading angle

Figure 5. Fringe Patterns, B Model, h = 3.18 mm

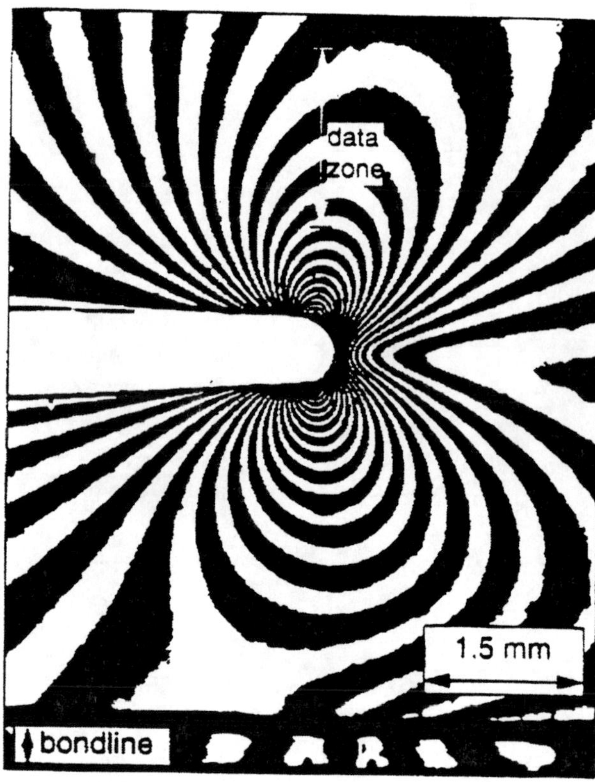

Figure 6 Normalized SIF Distributions for Bonded Homogeneous & Bimaterial A Models

but center slice results were always higher so only center slice results are reported. Fig. 9a is a thick slice near tip photo for the short crack in the bimaterial test (P3). The fringes are continuous across the upper bondline since the adhesive is chemically the same as the upper material (PSM9). However, there is a disturbance along the lower bondline where the material mismatch occurs. This pattern contrasts from the results for through cracks (Fig. 9b) This led to significantly different values of K_1 above and below the bondline for the part through crack. In order to determine which (if either) value of K_1 was correct in this case, a separate experiment (P7) was performed on a model consisting of materials A in Table II where there is no T_c mismatch and which had previously been found to yield good results (Smith, Finlayson and Liu, 1998) for through cracked specimens.

Results from the part through crack experiments are presented in normalized form in Table III.

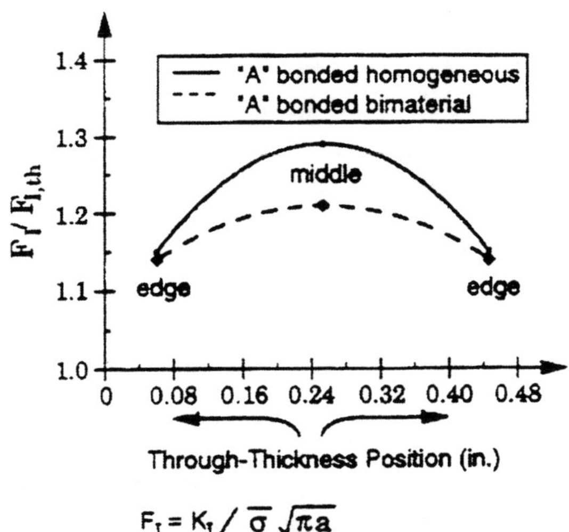

$$F_I = K_I / \overline{\sigma} \sqrt{\pi a}$$

$F_{I,th}$ = 2D Theoretical Estimate of K_I
Normalized by $\overline{\sigma} \sqrt{\pi a}$

Figure 7 Normalized SIF Distirbutions for Bonded
Homogeneous & Bimaterial A Models

Table III - Results for Part Through Cracks

Specimen	a/t	F_{1exp}	F_{1N}^{*}	F_{1N}^{**}
P1	0.39	1.06	1.18	1.11
P2	0.39	1.09	1.18	1.12
P3	0.41	1.21+	1.20	1.13
				F_{1N}^{****}
P4	0.69	1.54	1.42	1.50
P5	0.69	1.53	1.48	1.56
P6	0.69	1.53+	1.45	1.56
P7	0.69	1.47	1.43	1.42

+ - Average of F_1 values taken individually on each side of bondline

* - Values from Newman Theory (Newman, 1973) using average of three methods for computing equivalent semi-elliptic crack size, i.e., equal depth (a) and:

1) Equal areas ***

2) Equal major and minor axes *

3) Equal curvature **

$$F_1 = \frac{K_1}{\sigma \sqrt{\pi a}}$$

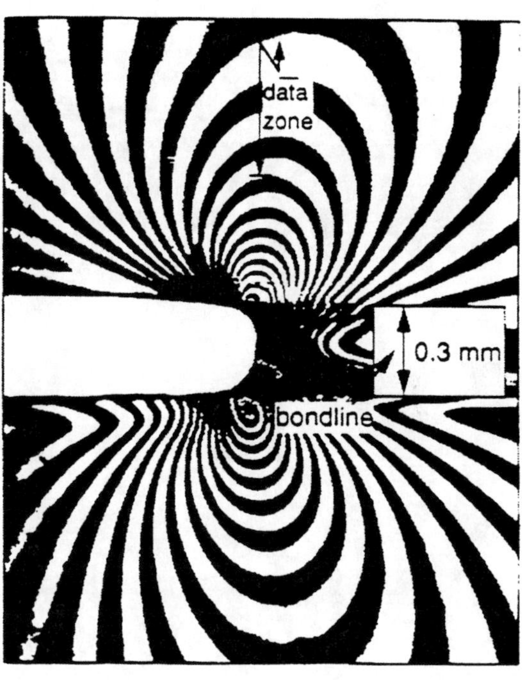

Figure 8. Fringe Pattern B Model, h = 0
(Through Thickness)

Figure 9. a) Fringe Pattern for Thick Slice (s = 7.77 mm) for Bimaterial Specimen P3. b) Fringe Pattern for Through Thickness Crack in Bimaterial Specimen (t = 12.7 mm).

Summary

Based upon a limited number of experiments, we conjecture that, for through thickness cracks:

i. There is negligible effect of modulus mismatch on SIF values for cracks in bondlines.

ii. There is a strong T_c effect for cracks parallel to and close to the bondlines of bimaterial specimens with T_c mismatch.

iii. For cracks <u>within</u> the bondline in bimaterial specimens with T_c mismatch, no T_c effect is detectable on SIF values.

iv. For the part through cracks the authors do not recommend the use of materials with differing T_c values. Nevertheless limited results suggest that such cracks are not nearly as severe as through cracks, for which such materials can be employed.

Acknowledgment

The authors wish to thank Air Force Research Laboratory through Raytheon STX Corporation for support of this project through subcontract No. 96-7055-LI467 with Raytheon STX Corporation. Special acknowledgement is also extended to Eric Finlayson who conducted all of the tests described herein.

References

Hutchinson, J. W. and Suo, Z., "Mixed Mode Cracking in Layered Materials," *Advances in Applied Mechanics*, Vol. 29, Academic Press, (1992).

Newman, J. C., Jr. "Fracture Analysis of Surface and Through Cracked Sheets and Plates," *Journal of Engineering Fracture Mechanics* Vol. 5, No. 3 pp. 667-689, (1973).

Smith, C. W., Finlayson, E. F. and Liu, C. T., "A Method for Evaluating Stress Intensity Factor Distribution for Cracks in Rocket Motor Bondlines," *Journal of Engineering Fracture Mechanics*, Vol.58, No. 1/2 pp. 97-105 (1997).

Smith, C. W. and Kobayashi, A. S., "Experimental Fracture Mechanics," Ch. 20 of <u>Handbook on Experimental Mechanics</u>, A. S. Kobayashi, Ed. VCH Publishers, (1987).

Smith, C. W., Finlayson, E. F. and Liu, C. T.,
"Effects on Cracks Within and Parallel to Bond-
lines by Stress Freezing Methods," <u>Experimental
Mechanics2 Advances in Testing and Design</u>,
I. M. Allison, Ed. A. A. Balkema, Brookfield,
pp. 1225-1230, (1998).

<u>Appendix</u>

<u>(Mode I Algorithm)</u>

Beginning with Griffith-Irwin Equations, we may
write, for Mode I, for the homogeneous case (Fig. A-
1).

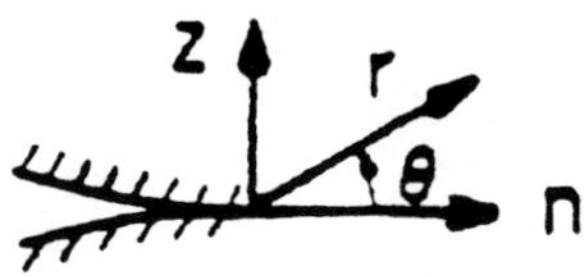

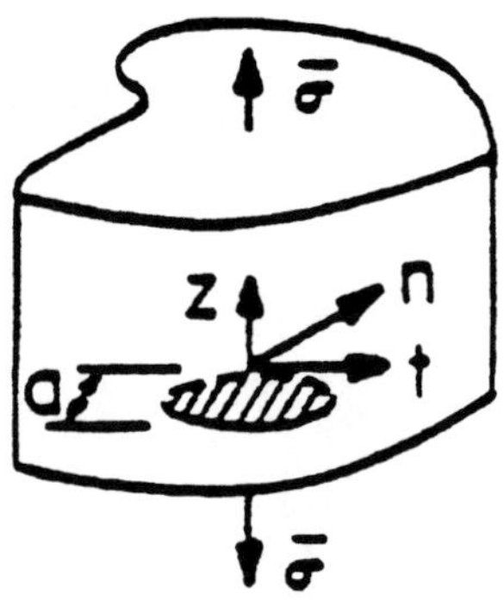

Fig. A-1 Mode I Notation

$$\sigma_{ij} = \frac{K_1}{(2\pi r)^{\frac{1}{2}}} f_{ij}(\theta) + \sigma_{ij}^0 \quad (i,j = n,z) \quad (1)$$

where

σ_{ij} = stress components

K_1 = Mode I SIF

$f_{ij} = f_{ij}(\theta)$

σ_{ij}^0 = non − singular stress components

Computing $(\tau_{nz})_{\max}$, truncating to same order as
Eq. (1) and evaluating along $\theta = \pi/2$, the direction of
maximum fringe gradient (Fig. A-2)

Fig. A-2 Mode I Fringes
(Pattern Unmultiplied)

$$(\tau_{nz})_{\max} = \frac{K_1}{(8\pi r)^{\frac{1}{2}}} + \tau^\circ = \frac{K_{AP}}{(8\pi r)^{\frac{1}{2}}} \quad (2)$$

where $\tau^\circ = f(\sigma_{ij}^0)$ and is constant over the data range,
K_{AP} = apparent SIF, $(\tau_{nz})_{\max}$ = maximum shear
stress in nz plane. Therefore

$$\frac{K_{AP}}{\bar{\sigma}(\pi a)^{\frac{1}{2}}} = \frac{K_1}{\bar{\sigma}(\pi a)^{\frac{1}{2}}} + \frac{\sqrt{8}\tau^\circ}{\bar{\sigma}} \left(\frac{r}{a}\right)^{\frac{1}{2}} \quad (3)$$

where (Fig. A-1) a = crack length, and $\bar{\sigma}$ = remote
normal stress, i.e., $\frac{K_{AP}}{\bar{\sigma}(\pi a)^{\frac{1}{2}}}$ vs. $\sqrt{\frac{r}{a}}$ is linear.

A typical plot of normalized K_{AP} vs. $\sqrt{r/a}$ for a
homogeneous specimen is shown in Fig. A-3.

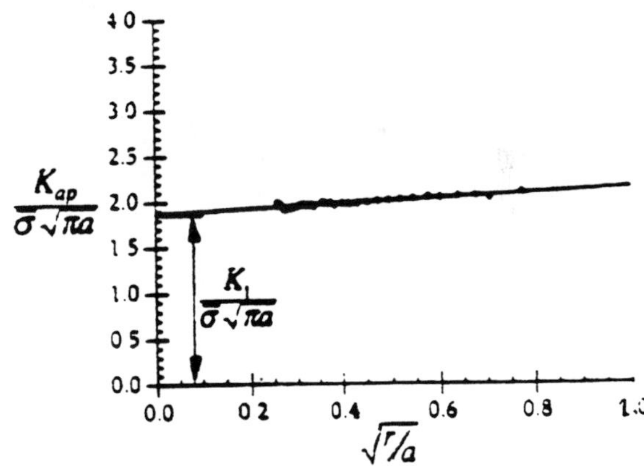

Fig. A-3 Middle Slice Data

Since from the Stress-Optic Law:

$$(\tau_{nz})_{\max} = \frac{nf}{2t} \begin{cases} n = \text{stress fringe order} \\ f = \text{material fringe value} \\ t = \text{slice thickness} \end{cases}$$

Therefore $K_{AP} = (\tau_{nz})_{\max}(8\pi r)^{1/2}$ is proportional
to n. The mixed mode algorithm follows a similar
procedure and is described in Smith and Kobayashi
(1987).

SHORT CARBON FIBERS COMPOSITES WITH HIGH FILLER CONTENTS RHEOLOGY AND MECHANICAL PROPERTIES

Y. SCHMITT, C. PAULICK, Y. BOUR, F.X. ROYER

Equipe "Matériaux Texturés et Céramiques"
Université de METZ
Laboratoire de Physique des Liquides et Interfaces
1 boulevard Arago, CP 97811
57078 METZ CEDEX 3
e-mail : royer@ipc.sciences.univ-metz.fr

ABSTRACT

The control of the quality of mixture based on very short carbon fibers and epoxyde resins leads to suitable mixture for molding of complex geometries. A gain in fluidity is obtained if the suspensions are treated by ultrasounds and simultaneously stirred under vacuum. Addition in a very small ratio of microbubbles in the mixture allows to obtain a viscosity less than those of the matrix alone. For many polymer materials the gain of fluidity can be of 20 to 25% with size and concentration of the microspheres thoroughly chosen. A certain number of new resins is developped to elaborate composite materials with specific mechanical properties close to standard aluminium. Tensile test an ultimate stress are used to quantify the improvements of the mechanical properties. Fillers concentrations up to 30 % are obtained.

INTRODUCTION

Short fiber composite materials are known since some time. Although they may in general be transformed by molding without to much difficulties, there are applications for which either undesirable results occur due to partial orientation of the fibers or the number of pieces to be fabricated does hardly justify the expense for the mold. In both cases it seems to be interesting to look for mixtures of lower viscosity in the liquid phase but yielding materials of a comparable strength.

A second point of interest is the resistance against corrosion of the material. It can be noted that the highly corrosion resistant materials based on polypropylene or polyethylene are of rather poor mechanical performance, whereas polyamide based composite materials of good strength are not as corrosion resistant as one might wish.

This publication reports the work on short fiber composite materials based on epoxyde resins with milled carbon fibers as reinforcing particles [1, 2]. In the first part we discuss the rheological behavior of mixtures ranging from 0 to 30 % volumic fraction of fillers and the influence of small quantities of glassy microspheres on the viscosity of the mixtures. Due to measurements with several epoxyde resins having viscosities between 300 mPa·s and approximately 3000 mPa·s, it can be shown that the flow is behaving differently for the mixtures based on a low viscosity matrix compared to those mixtures which are prepared with resins of higher viscosity. Below a matrix viscosity of 800 mPa·s no significant variation of the viscosity of the composite mixtures is found, but if a matrix with higher viscosity is used a pronounced increase of the viscosity of the mixtures will be observed.

In the second part the mechanical strength (limit at break) of the materials is analyzed with regard to properties of the matrix and fiber length distribution. As well is discussed the influence of the orientation of the fibers which may either result from the molding process or be obtained by a special treatment with magnetic fields [3, 4]. In order to obtain good mechanical strength the fibers should have a certain length at least above what is called the critical length l_c. The formulae of the critical lenght retained for this work is given by Hsueh [5] :

$$\frac{l_c}{r} = 4.7\sqrt{\frac{E_f}{E_m}} \qquad (1)$$

where r is the radius of fiber , E_f and E_m respectively the elastic moduli of the fibers and the matrix.

As the interactions between fibers in a liquid mixture are rapidly increasing with the length of the fibers [6, 7], thus increasing as well the viscosity, it is important to use an epoxy matrix with a Young modulus as high as possible to reduce the critical length of the fibers. To our knowledge, commercially available epoxyde resins which have an elevated Young modulus (4 to 5 GPa) are rather

viscous (> 2500 mPa·s) and therefore not of much interest for the mixtures to be made. This conclusion triggered the development of a new family of resins which are as well very fluid (300 to 400 mPa·s) and quite stiff (Young modulus > 4 GPa) (table 1, 2).

Table 1 : Origin and codification of the resin/hardener systems

n°	resin	manufacturor
1	ARC-40/30	ARC
2	ARC-40/33	ARC
3	ARC-40/39	ARC
4	ARC-40/32	ARC
5	ARC-40/30	ARC
6	ARC-24/32	ARC
7	RE1820/DE1825	MYBOND
8	LY564/HY2954	CIBA GEYGY
9	ARC-12/15	ARC
10	AY103/HY991	CIBA GEYGY
11	AY103/HY991	CIBA GEYGY
12	AY103/HY991	CIBA GEYGY
13	ARC-40/34	ARC
14	ARC-40/35	ARC

Table 2 : Elastic modulus, viscosity at 20 °C and critical length calculated with eq. 1 for the twelve resins referenced in Table 1.

n°	modulus E_m (MPa)	viscosity (Pa.s)	critical length l_c (µm)
9	1600	1,2	197,2
10	2100	6,2	172,2
8	2600	1	154,7
7	3100	0,8	141,7
6	3300	0,6	137,3
3	3600	0,38	131,5
4	3900	0,4	126,3
13	4010	0,41	124,6
14	4150	0,42	122,5
5	4300	0,42	120,3
2	4600	0,35	116,3
1	4900	0,3	112,7

The samples were characterized with traction, compression and flexion measurements and the orientation of the fibers is obtained by digital treatment of microscope images.

EXPERIMENTS

Viscosity measurements

Before measuring the charged resins each epoxyde system was characterized regarding the viscosity vs. time curve. This aims to obtain the period before the onset of polymerization when the viscosity of the matrix was not yet altered. Typically this time is about 30 minutes for the systems studied.

If not stated otherwise the measurements are made with a rotating cone-plate instrument ("Rheotest", Prüfgeräte Meidingen, former R.D.A.) at a fixed shear rate which can be varied from 0.50 to 900 s^{-1}. The precision of the results is typically of 3%, this error being as well the standard deviation for a series of five measurements and the error given by the company for this instrument.

The fibers used in these experiments are milled T300 PAN fibers which were purchased from Apply Carbon Ltd. (France). The fiber length distribution was determined by numerical treatment of microscope images ("Visilog5" from Noesis, France) for a sample of approximately 6000 fibers and is shown in figure 1. As it can be seen below, this type of fibers, presenting a mean length of approximately 160 µm, is well suited to make the composite materials.

Epoxyde resins used are either commercial qualities from CIBA-GEIGY and MYDRIN-LAMBIOTTE or samples from a newly developed series of resins from ARC Ltd., the references are given in table 1. They are used as prescribed by the respective manufacturers. Table 1 summarizes the main properties of the studied resins.

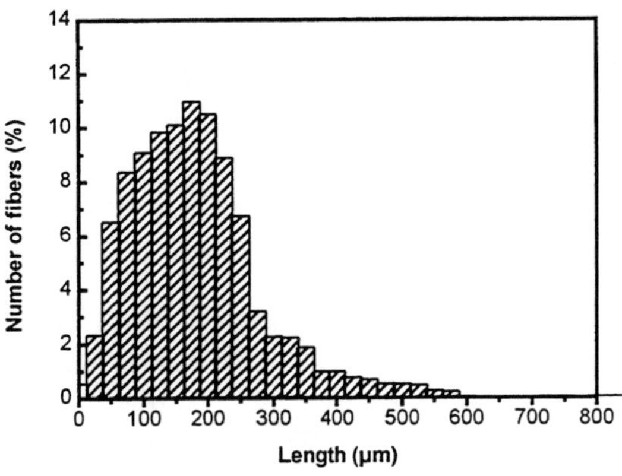

Figure 1 : fiber length distribution of T300-mld short fibers

In the first experiment the mixtures have been prepared in a little mixer for some minutes and are rapidly transferred to the viscometer. Figure 2 shows the viscosity of twelve resins charged with 5, 15 and 30 % in volume of carbon fibers at a shear rate of 100 s^{-1}. It can be seen, that there is almost no increase of the viscosity of the mixtures for the first group of resins and the composites became more viscous only when the epoxyde matrix has a viscosity above 800 mPa·s. These results suggest that there are at least two contributions leading viscous dissipation. At low viscosities the dominant contribution comes from direct and limited interactions between fibers whereas above the threshold the matrix will "help" more and more the formation of bigger agglomerates of fibers thus increasing rapidly the viscosity. It should be noted that these observations are independent of the fiber content. To our knowledge there is no theoretical explanation for this phenomenon today.

It should be added that the same behavior can be found at different shear rates, the value of 100 s^{-1} chosen represents an intermediate range of flow where the orientations of the fibers are neither completely random nor fully aligned with the shear gradient. As here we do not want to treat the transition from the random orientation to the partially aligned state which occurs in the very beginning of the measurement and which leads to a peak of the viscosity [8] that disappears with a characteristic relaxation time of approximately 1 minute. The measurements presented are limited to the period after the viscosity has stabilized. For one resin of the first series the viscosity is given in function of the shear rate and the fiber content in figure 3. As expected the increasing fiber content leads to an increase in viscosity and shear thinning behavior which is quite common and well described in literature [9, 10, 11, 12].

In a second experiment the influence of several treatments of the mixtures is studied. In addition to mechanical stirring the mixtures are exposed to ultrasound and other mixtures are tested which contain in addition to the fibers 1 % of glassy microspheres. The variation of the relative viscosity is plotted in figure 4. The relative viscosity is defined as the ratio of the viscosity of the treated mixture by the viscosity of the mixture with any treatment.

Either ultrasound or addition of microbubbles or both provoke a reduction of the viscosity. This reduction is most pronounced at low shear rates and comes from a partial destruction of the agglomerates of fibers. At higher shear rates the measurement itself will destroy the agglomerates and align the fibers as discussed above and therefore the gain in fluidity is less pronounced with the ultrasonic treatment. An unexpected and important reduction of the viscosity is found with the addition of the microspheres.

The results of viscosity measurements are shown in figure 5 which demonstrate that in spite of what could be expected 1 % in volume of microspheres may lower the viscosity of a charged resin 25% below the value of the uncharged one for the shear thinning part of flow.

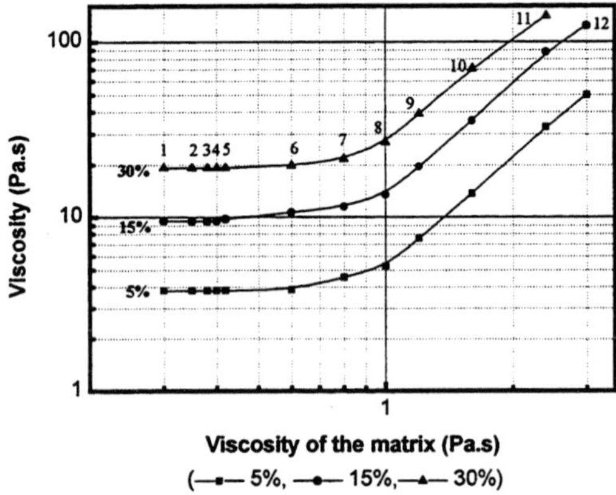

Figure 2 : Composite viscosity versus matrix viscosity

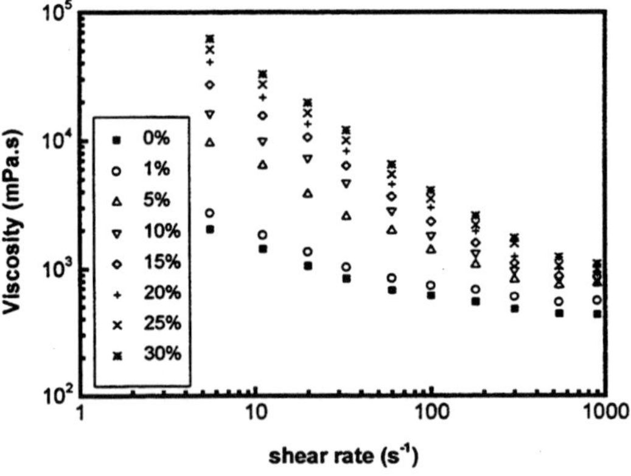

Figure 3 : Viscosity of the resin ARC-40/32 as a function of shear rate and fiber content

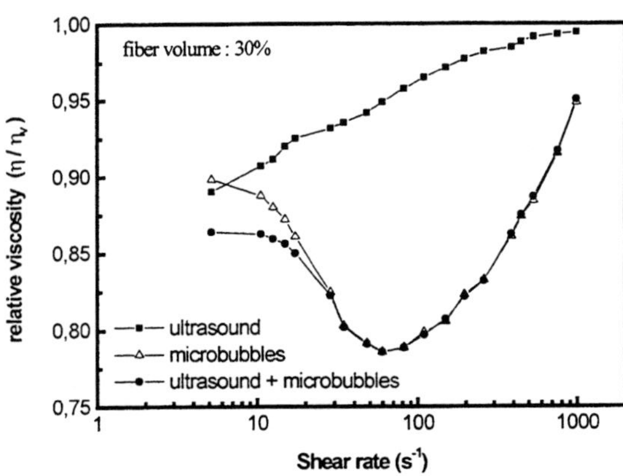

Figure 4 : Influence of several treatments on the composite mixture

This observation is clearly in contrast to earlier work, especially the theory of Einstein [13] and later works based on the same approach [14, 15, 16], but to our knowledge it was never investigated systematically what influence particles may have on the viscous behavior of a liquid which is already pseudoplastic. One can conclude from the data of figure 5 that the reduction of the viscosity is related to the form of the particles because platelets and fibers which are investigated as well do not produce any gain in fluidity.

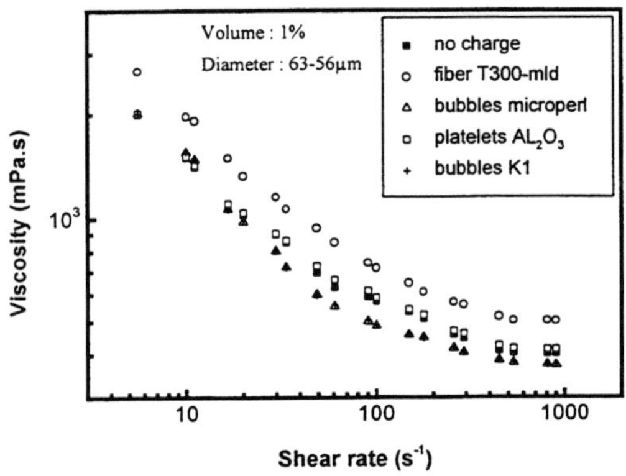

Figure 5 : Influence of the aspect ration of the particule on the viscosity.

Notice that K1 and Microperl are both glassy microbubbles respectively empty and full produced by 3M and SOVITEC.

A more detailed analysis of this new rheologic behavior will be published elsewhere [17]. Please note that patent applications are pending for applications related to particular concentrations of microspheres. In general 1 % in volume of microbubbles of the good size added to the mixtures will compensate the viscosity increase due to an additional amount of 5 % in volume of fibers. The latter mixture permits a gain in mechanical properties up to 40% compared to the mixture without microbubbles and lesser fiber content.

Mechanical measurements

The same mixtures which were investigated for their rheological properties are molded into plates of approximately 100 x 200 mm^2 and a thickness of about 6 mm. Molding takes place under ambient pressure in open forms. Samples for testing are machined from these plates. The strength at break is determined in an uniaxiale tensile test with dumbbell specimens according to the norm ISO R527 (1966). For compression the testing is done following the recommendations of the norm ASTM D495-69 and flexion experiments are made following the norm ISO 167(1975). The experiments are always continued until complete destruction of the sample, which at the speed of 1mm/minute in general happens instantaneously.

As in this paper we want rather to present a certain number of data together with some general trends the quite sophisticated models found in the literature [18, 19, 20, 21] are not treated here. A more detailed interpretation of the results and a comparison with existing theories will be published elsewhere [22].

A good idea about the important parameters for these materials may be obtained from the figures 6a to 6c. Fibers concentration and fibers length distribution are similar for the three materials but the stress-strain curves of the epoxyde resins used as matrix are quite different. First it can be noted, that the deformation at break does not depend of the matrix. The composite based on an Araldite® epoxyde resin which, as pure resins deforms 3.5%, has about 1% strain at break as well the composite based on the specially developed resin ARC 40/33 exhibiting no more than 2% of deformation and almost no plasticity. Only composite materials with less than 15% volumic fraction of fibers show deformations greater than 1.2%. These results prove that for these composites for which a good adhesion between the fibers and the resin may be assumed there is no interest to have an important plastic deformation of the matrix. Regarding the tensile yield stress it can be seen, that the most important factor is the Young modulus of the matrix. A 50 % increase in the Young modulus doubles the tensile yield stress (comparison between figure 6b and 6c).

The figure 7 shows the tensile stress at break for a series of composites with 30 % fiber content compared to that of the pure resin in function of the Young modulus of each resin. No correlation with the tensile strength at break of the matrix can be found with the exception of the last system which was made with a very brittle epoxy of poor mechanical strength. It is found that the strain at break is about 1% for all but the last composite, the latter with a matrix that deforms less than the fibers is meant to check the validity of the assumptions and can not be directly compared with the other systems. Because all composites are made up of the same fibers but of a different matrix they will differ also regarding the fraction of fibers above and below the critical fiber length l_c (eq. 1).

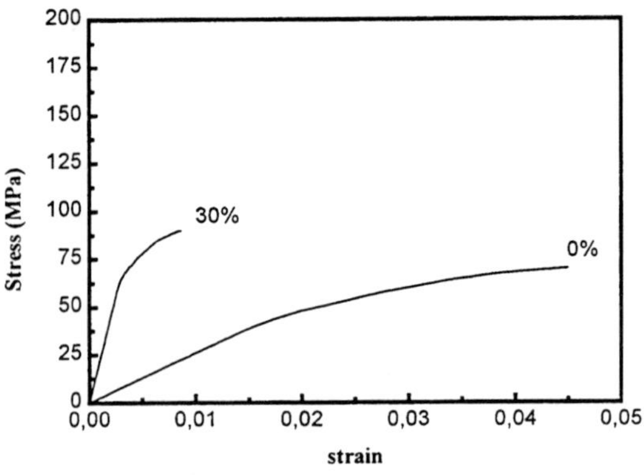

a) Araldite LY564/HY2954

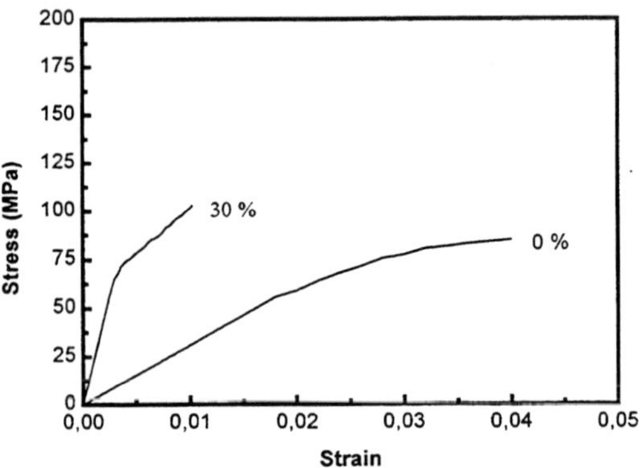

b) Mybond RE1820/DE1825

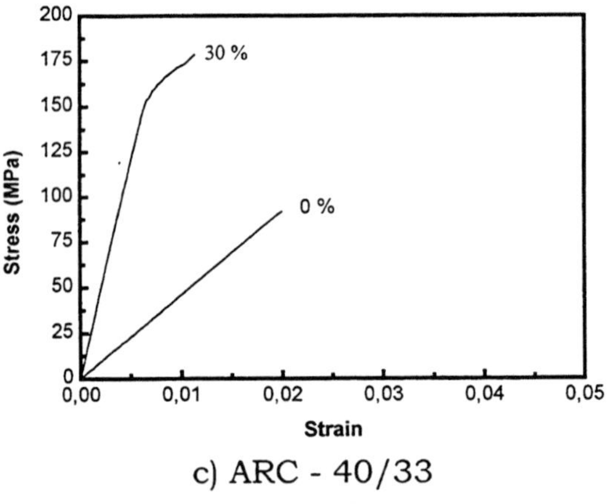

c) ARC - 40/33

Figure 6 : Influence of the matrix on the mechanical properties of the final composite

The fact that there is a considerable fraction of the fibers below the critical length for the less rigid resins explains well why only a little gain in

22

mechanical resistance. Nevertheless the increase in ultimate strength for the resins with a Young modulus between 3900 MPa and 4600 MPa can not be explained alone by the variation of the critical lentgth l_c.

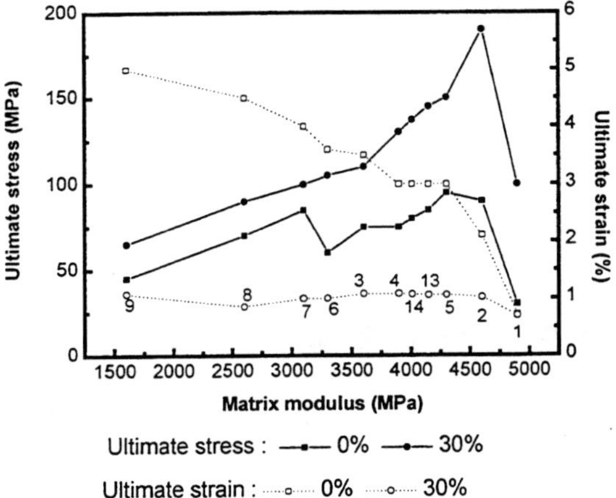

Figure 7 : Influence of the matrix modulus on the ultimate stress for 30% T300-mld composites

CONCLUSION

Physical treatments such as ultrasounds or stirring under vacuum have a significant influence on the rheological properties and participate to decrease the viscosity of the mixture short carbon fibers and resins.
Adding a small ratio of glassy microbubbles also contribute to decrease the viscosity of the molding mixtures and of course of the porosity.

The critical length of the fibers roughly influence the properties of the final composite in connection with the Young modulus of the matrix. An increase of the fiber length does not automatically leads to a gain on the mechanical properties. If this length is close to the critical length the elastic modulus of the resins becomes a predominant factor for the ultimate properties of the composite.

For the carbon fibers composite the mechanical properties increase with the filler aspect ratio.

Vacuum+ultrasounds+micro-bubbles permit to reach ratios of about 35% fibers in the composites.
The improved techniques describe in this paper are applied for the production of a new cellular composite.

REFERENCES

[1] SCHMITT Y. (1999). "Approches rhéologique et mécanique des matériaux composites à fibres courtes", thesis of the University of METZ.

[2] BOUR Y. (1999). "Matériaux composites à fibres courtes: relation entre facteur d'orientation et propriétés mécaniques et thermiques", thesis of the University of METZ.

[3] SCHMITT Y., PAULICK P., ROYER FX, GASSER J.G. (1996) "Magnetic field induced orientational order of conductive fibers in non-conductive liquids", Journal of Non Crystalline Solids, vol.205-207, p135-138.

[4] BOUR Y., PAULICK C., SCHMITT Y., ROYER F.X. (1998). "Elaboration et caractérisation de matériaux composites orientés sous champs magnétiques", C.R. JNC 11, Arcachon, France, vol.1, p.127-130.

[5] HSUEH C.-H. (1995). "Modeling of elastic stress transfer in fiber-reinforced composites". Trends in Polymer Science, vol.3, n°10, p.336.

[6] BIBBO M.A., DINH S.M., ARMSTRONG R.C. (1985). "Shear flow properties of semiconcentrated fiber suspensions". Journal of Rheology, vol.29, p.905.

[7] DINH S.M., ARMSTRONG R.C. (1984). "A rheological equation of state for semiconcentrated fiber suspensions". Journal of Rheology, vol.28, p.207.

[8] KAMAL M.R., MUTEL A.T. (1985). "Rheological properties of suspension in newtonian and non newtonian fluid". Journal of Polymer Engineering, vol.5, n°4, p.293.

[9] BATCHELOR G.K. (1974). "Transport properties of two-phase materials with random structure". Annual Review of Fluid Mechanics, vol.6, p.227.

[10] HAND G.L. (1962). "A theory of anisotropic fluids". Journal of Fluid Mechanics, vol.13, p.33.

[11] KAMAL M.R., MUTEL A.T. (1989). "The prediction of flow and orientation behavior of short fiber reinforced melts in simple flow systems ". Polymer Composites, vol.10, p.343.

[12] MUTEL A.T., KAMAL M.R. (1984). "The effect of glass fibers on the rheological behavior of prolypropylene melts between rotating parallel plates". Polymer Composites, vol.5, p.29.

[13] EINSTEIN A. (1956). "Investigation on the theory of the brownian movement". New York, Dover.

[14] BATCHELOR G.K., GREEN J.T. (1972). "The determination of the bulk stress in a suspension of spherical particles to order c²". Journal of Fluid Mechanics, vol.56, n°3, p.401.

[15] BRADY J.F., BOSSIS G. (1985). "The rheology of concentrated suspensions of spheres in simple shear flow by numerical simulation". Journal of Fluid Mechanics, vol.155, p.105.

[16] ROSCOE R. (1952). "The viscosity of suspensions of rigid spheres". British Journal of Applied Physics, vol.3, p.267.

[17] SCHMITT Y., PAULICK P., BOUR Y., ROYER FX. (1999). "Influence of the addition of microbubbles on the rheology of polymer mistures". To be published.

[18] AHMED S., JONES F.R. (1990). "A review of particulate reinforcement theories for polymer composites". Journal of Materials Science, vol.25, p.4933.

[19] ESHELBY J.D. (1961). "Elastic inclusions and inhomogeneities". Progress in Solid Mechanics, Sueddon I.N. & Hill R., North Holland Publishing.

[20] LAUKE B. (1990). "Theorical considerations of toughness of short-fiber-reinforced thermoplastics". Polymer - Plastics Technology Engineering, vol.29, n°7 et 8, p.607.

[21] BERRYMAN J.G., BERGE P.A. (1995). "Critique of two explicite schemes for estimating elastic properties of multiphase composites". Mechanic of Materials, vol.2, p.149.

[22] SCHMITT Y., PAULICK P., BOUR Y., ROYER FX. (1999). "Short fibers composite : comparison between experimental data and results of different behaviour models".
To be published.

Effect of Thermal Residual Stresses
on Polymer/Metal Interfacial Adhesion

Qizhou Yao and Jianmin Qu
G. W. Woodruff School of Mechanical Engineering
and
Packaging Research Center
Georgia Institute of Technology
Atlanta, GA 30332-0405

Abstract

In this study, the apparent fracture toughness of the interfaces of several epoxy-based polymeric adhesives and metal (aluminum) substrate is experimentally measured. Double layer specimens with initial interfacial cracks are made for four-point bending tests. Thermal residual stresses exist on the interface due to the coefficient of thermal expansion (CTE) mismatch between the underfill and aluminum. Silica fillers are used to modify the CTE of the epoxy-based adhesives so that various levels of interface thermal residual stresses are achieved. Finite element analysis is also performed to quantify the effects of CTE mismatch as well as the elastic mismatch across the interface. It is found that the apparent interfacial toughness is significantly affected by the thermal residual stress, while the effect of elastic mismatch is negligible. In general thermal residual stress undermines the resistance to an interfacial crack. In some cases the residual stress is sufficient to result in adhesive and/or cohesive failure.

Introduction

Thermal residual stresses in electronic assemblies result from the thermal and stiffness mismatch of bonded materials. During the manufacturing processes and service operations, significant temperature cycling is present in the microelectronic devices, and the free-edges of the bonding interfaces in such structures suffer high stress gradients, which eventually cause cracking of the layered structures or interfacial delamination[1]. Thermally induced stresses in laminated packaging structures have been studied with or without considering viscoelastic properties of polymeric materials[2, 3], or creep deformation and stress relaxation of polymeric materials[4]. It is perceivable that the residual stress plays an important role in the reliability of thin film interconnects. In some cases, the thermal induced stress itself can exceed metallization yield strength[4]. These residual stresses need to be taken into account when conducting reliability and life prediction analyses.

Fracture mechanics approaches have been used to study reliability problems, in particular, adhesion related failure in electronic packages. The J-integral criterion is mostly used to characterize both polymeric materials and interfacial adhesion. However, in the past the effect of thermal residual stresses on the interfacial adhesion was generally not considered[5-7]. Consequently, the interfacial toughness measured experimentally is not an intrinsic bimaterial property. Instead it is an apparent parameter which depends on the thermal residual stresses exist in the bimaterial assembly. To quantify the intrinsic toughness of bimaterial interfaces, the effect of thermal residual stress on the apparent interfacial toughness needs to be identified. This is particularly important for interfacial adhesion characterization, since the CTE mismatch across the interface is often significant when polymeric/metal interfaces are considered.

In this work, the apparent fracture toughness of underfill/aluminum interfaces is obtained through interfacial fracture tests. The finite element analysis (FEA) is conducted to calculate the thermal induced J-integral associated with the interfacial crack. Results of the FEA also provide the basic tool for studying the effect of elastic mismatch on the interfacial toughness.

Experiment Program

Adhesive Preparation: A cycloaliphatic epoxy based underfill adhesive is selected as the base adhesive. To alter the coefficient of thermal expansion (CTE) of the adhesive, 5 - 10 μm silica power is used as fillers and four alterations of the underfill adhesives are made with 15%, 30%, 45%, and 60% (wt%) silica fillers. These adhesives are denoted as Base, 15F, 30F, 45F, and 60F. The CTE and elastic modulus of these adhesives are measured and listed in Table 1.

Table 1. Properties of adhesives

Adhesive	Base	15F	30F	45F	60F
Filler wt.%	0	15	30	45	60
CTE ($10^{-6}/^{0}$C)	83.2	78.4	70.4	56.0	29.6
E (GPa)	3.19	3.71	4.26	4.97	5.83

Sample Preparation: Aluminum strips of 75 x 9.5 x 1.60 mm are prepared as substrate. Adhesives are applied on the top of aluminum strips. Specimens with two adhesive layer thickness, 0.80 mm and 3.20 mm, are made. The aluminum surface at the interface side is polished using grid 800 sanding paper to ensure flat and smooth surface. The adhesive thickness non-uniformity for each specimen is controlled to be less than 0.03 mm. Specimen curing profile is 250 ^{0}C/15min in air. Symmetric pre-cracks are created in the interface by using a molding compound release agent.

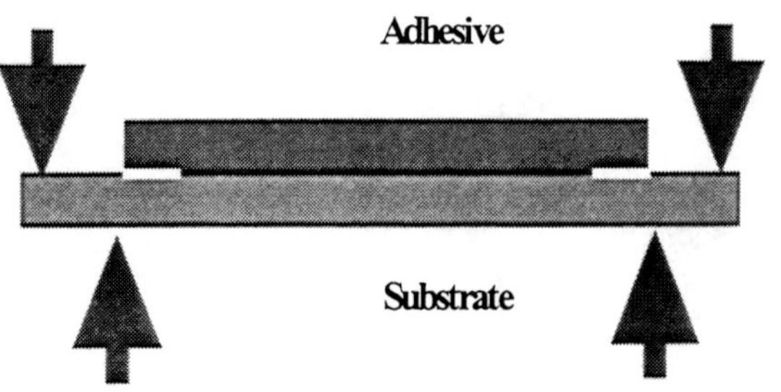

Figure 1. Double-layer specimen

The surface preparation of aluminum substrate prior to the underfill application is based on a method proposed by Wong and McBride.[8] Cleaning steps include: 1) 5 min. soak in terpene organic solvent; 2) 5 min. soak in terpene during ultrasonic cleaning; 3) 5 min. soak in isopropyl alcohol; 4) 5 min. soak in isopropyl alcohol during ultrasonic cleaning; 5) 3 rinses in deionized water; 6) 5 min. soak in deionized water during ultrasonic cleaning; 7) rinse in deionized water; 8) bake in clean oven at 120 ^{0}C for 30 min. under vacuum; and 9) 5 min. UV/ozone treatment at 50 ^{0}C.

<u>Fracture Test:</u> Specimens are subjected to four-point bending after the adhesives are cured, and the critical load at which the crack starts to propagate is recorded from the load-displacement curve. Figure 2 shows a typical load-displacement curve and the critical load is the point corresponding to initial load drop.

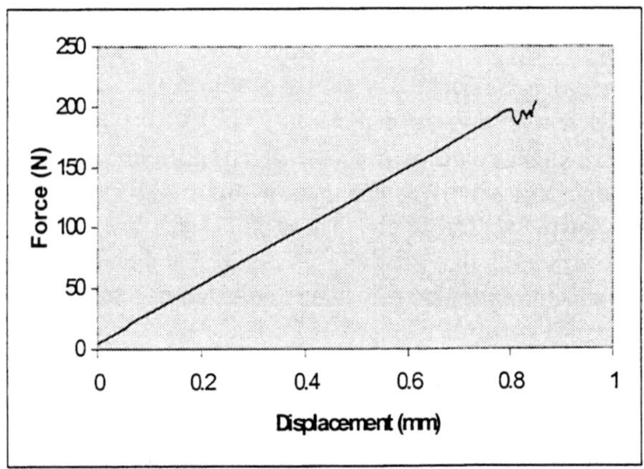

Figure 2, Typical load-displacement response of the four-point bending specimens.

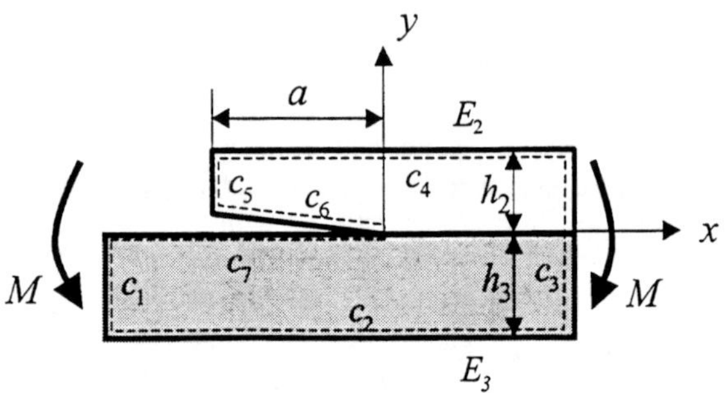

Figure 3, Free body diagram of a specimen with an interface crack.

For the crack tip shown in Figure 3, the singular stress field ahead of the crack tip is given by

$$\sigma_x + i\sigma_{xy} = \frac{K}{\sqrt{2\pi x}} x^{i\varepsilon} \;, \qquad (1)$$

where K is the complex stress intensity factor

$$K = K_1 + iK_2 \;, \qquad (2)$$

and ε is the bimaterial constant that depends on elastic constants of the two materials involved.

In (2), K_1 and K_2 can be viewed as the crack-tip Mode I and Mode II stress intensity factors, respectively, although they do not carry the same physical meaning as their counterparts in the case of cracks in homogeneous media. This is because that K_1 and K_2 have a strange, material dependant dimension and they are intrinsically coupled.

Since $x^{i\varepsilon} = \cos(\varepsilon \ln x) + i \sin(\varepsilon \ln x)$, it then follows from (1) that the crack-tip stress field is oscillatory. This pathological behavior has been a stumbling block for the development of interfacial fracture mechanics. Since K is a complex number, one may define its amplitude and phase by introducing a characteristic length parameter L so that

$$KL^{i\varepsilon} = |K|e^{i\psi} \;, \qquad (3)$$

where ψ is called the phase angle of the complex quantity $KL^{i\varepsilon}$, giving $|L^{i\varepsilon}| = 1$ and $|KL^{i\varepsilon}| = |K|$. The choice for L is arbitrary[10], but must be chosen as a fixed length and reported since it affects the meaning of ψ. Usually, it is convenient to choose a geometry dimension (crack length, layer thickness, etc.) as the characteristic length L when stress analysis is concerned. On the other hand, if the focus is on material failure and microstructural analysis of crack-tip field damage, a length related to material microstructure (e.g., grain size) may be chosen as the characteristic length L. The characteristic length, L, used throughout the calculation is the length of the specimen, e.g. 76.2 mm.

The energy release rate G, which is the amount of energy released per unit area of an interface to decohere, describes the loading amplitude and is given by

$$G = \frac{|K|^2}{E^* \cosh^2(\pi\varepsilon)} \;, \qquad (4)$$

where

$$\frac{2}{E^*} = \frac{1}{\overline{E}_1} + \frac{1}{\overline{E}_2} \;, \qquad (5)$$

and $E'_n = 2\mu_n /(1 - \upsilon_n)$ is the plane strain modulus.

At a given phase angle ψ, the maximum loading amplitude G that an interface can sustain without decohesion is called the toughness of the interface at this phase, G_c. By measuring the critical load, P_c, at which fracture occurs in a specimen for a particular phase angle ψ, the value for the critical stress intensity factor, K_c, and thus the value for the interfacial fracture toughness, G_c, can be obtained. An interfacial fracture toughness curve is generated by determining the critical load P_c for different values of the phase angle ψ, and then calculating the corresponding value G_c.

Finite Element Modeling

Based on plane strain assumption two dimensional finite element models are constructed to calculate the critical strain energy release rate, G_c, corresponding the experimentally measured critical load. The G_c obtained here is considered as FEM interfacial toughness associated with the interface crack. The FE calculation is carried out by ABAQUS Standard version 5.7. Several FE models with different adhesive and aluminum thickness are employed to explore the interfacial toughness dependency on mode mixity over certain range. Figure 4 shows the meshes and deformation of one such FE model. Note that the mesh around the crack-tip area is refined to achieve improved G_c estimation.

Since thermal residual stress results from both CTE and stiffness mismatch across the interface, there is a need to identify which one affects the interfacial toughness more than the other. In the FE models, artificial elastic modulus is assigned to adhesives with same CTE, and the energy release rates of different cases are compared.

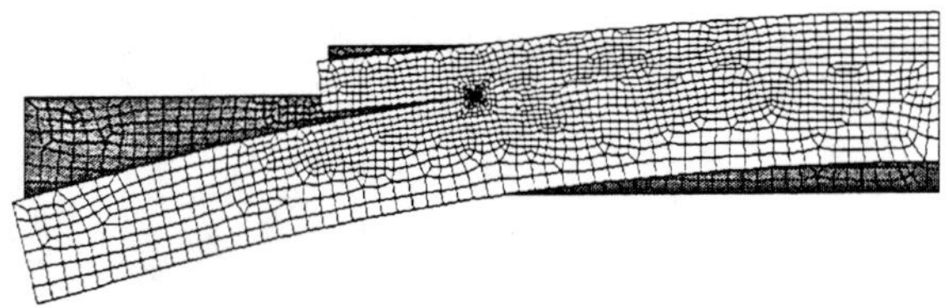

Figure 4, Meshes and deformation of one-half specimen subject to four-point bending.

Results and Discussion

Figure 5 shows the measured interfacial toughness for Base adhesive and four silica-filled adhesives. To assure the same basis of comparison, these experimental data shown in Figure 5 are chosen such that the corresponding mode mixity of each specimen is in the neighborhood of 47^0. It is seen that specimen 30F has the highest apparent interfacial fracture toughness. Notice that the thermal residual stresses decreases as the filler content increases. Therefore, it seems that there is an optimal value for the underfill CTE at which the apparent interfacial toughness is the best.

However, one must understand that higher filler content decreases the CTE, but also increases the modules, see Table 1. That means the effects shown in Figure 5 reflect not only the residual stresses, but also the elastic mismatch. Ideally, to investigate the residual stress effects, one should find several materials with same modulus, but different CTE values. Practically, this is almost impossible. Therefore, the

results shown in Figure 5 represent the combined effects of both modulus change and CTE change.

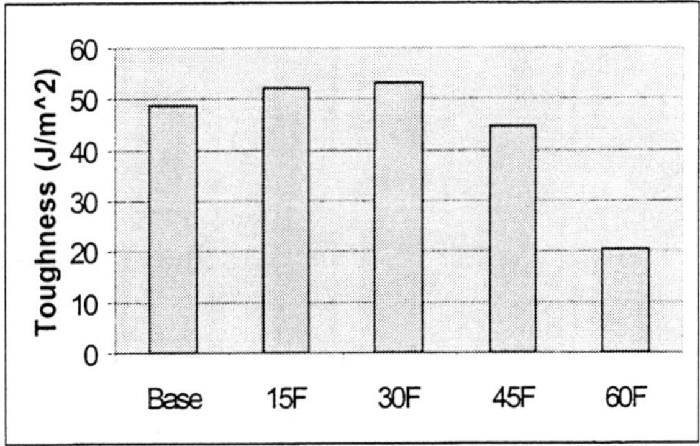

Figure 5, Apparent toughness of underfill/Al interfaces

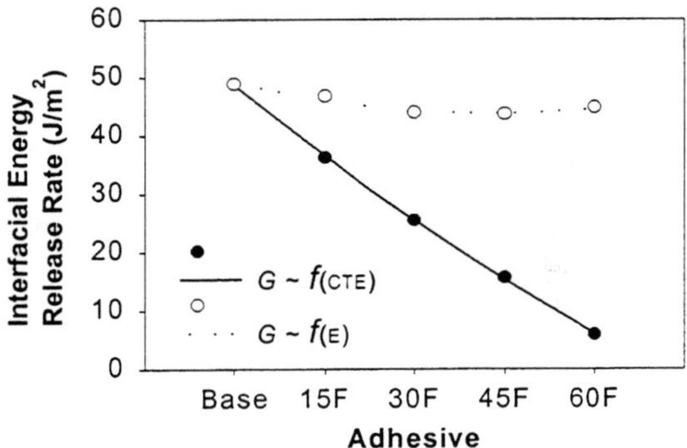

Figure 6, Interfacial energy release rate as functions of CTE and elastic modulus

To separate these two effects, two types of computations can be performed. First, keep the modulus unchanged and vary the CTE. This will qualitatively show how the interfacial energy release rate changes as a function of adhesive CTE only. Second, one can keep the CTE constant and vary the modulus. This will extract the effect caused by the modulus variations. The results these computations are shown in Figure 6. Clearly, the effect of CTE is much more significant, while the effect of modulus variation may be negligible.

Conclusions

The apparent fracture toughness of the interfaces of several epoxy based underfill adhesives and aluminum substrate is experimentally measured. The measurements are correlated to different levels of interface thermal residual stresses resulted from the Coefficient of Thermal Expansion (CTE) mismatch of underfill and aluminum. It is found that thermal residual stress plays considerably detrimental role on the apparent adhesion interfacial toughness. The effect of the elastic modulus changes associated with the change of CTE may be insignificant.

References

1. Z.Q. Jiang *et al.*, "Thermal Stresses in Layered Electronic Assemblies." ASME *J. of Electronic Packaging,* Vol 119(1997), pp.127-132.

2. E. Suhir, "An Approximate Analysis of Stresses in Multilayered Elastic Thin Films," ASME J. of Appl. Mechanics, Vol. 110 (1988), pp.143.

3. M. Pecht, *et al.*, "To Cut or Not to Cut: A Thermomechanical Stress Analysis of Polyimide Thin Film on Ceramic Structures," IEEE *Transactions on Components, Packaging, and Manufacturing Technology,* Part B, Vol 18, 1(1995), pp.150.

4. S.X. Wu, *et al.*, "A Consititutive Model of Polyimide Films and Its Instgration with Finite Element Analysis for Residual Stress Prediction in Thin Film Interconnects," ASME EEP-Vol. 19-2(1997), Advances in Electronic Packaging, pp.1285-1290.

5. P.D. Brandenburger and R.A. Pearson, "Mixed Mode Fracture of Organic Chip Attachment Adhesives," ASME EEP-Vol. 11/MD-Vol. 64(1995), Application of Fracture Mechanics in Electronic Packaging and Materials, pp. 179-185.

6. V. Sundararaman, "New Test Methods For Determining Fracture Toughness As A Function of Mode Mix For Bimaterial Interfaces," *ibid*, pp. 141-154.

7. M.B. Vincent, L. Meyers, and C.P. Wong, "Enhancement of Underfill Adhesion to Die and Substrate by Use of Silane Additives," *4th International Symposium and Exhibition on Advanced Packaging Materials.* pp. 49-52.

8. C.P. Wong and R. McBride, "Preencapsulation Cleaning Methods and Control for Microelectronics Packaging," *IEEE Transactions on Components, Packaging, and Manufacturing Technology-Part A*, Vol.17-4, pp. 542-552, (1994).

9. Dundurs, "Edge-bonded dissimilar orthogonal elastic wedges". *J. Appl. Mech.* 36 (1969), pp 650-652.

10. J.W. Hutchinson and Z. Suo, "Mixed Mode Cracking in Layered Materials," in Advances in Applied Mechanics, Vol 29, ed. By Hutchinson J.W. and Wu, T.Y., Academic Press, New York, 1991.

11. Q. Yao and J. Qu, "Interfacial Adhesion," manuscript in preparation (1999).

12. Kuhl and J. Qu, "A Technique for Interfacial Toughness Measurement," *J. Electronic Packaging*, submitted.

MD-Vol. 88, Polymeric Systems
ASME 1999

In-situ Continuous Process for Bonding
Glass-Fiber Reinforced Polypropylene to Wood

Karthik Ramani*
Mechanical Engineering, Purdue
University,
West Lafayette, IN 47907-1288, USA
765 494-5725

Michael Smith
1427 Lyndon #C
South Pasadena, CA 91030
626 799-2703
thesmith@alumni.caltech.edu

Heming Dai
Mechanical Engineering, Purdue University
West Lafayette, IN 47907-1288, USA
765 494-5649

ABSTRACT

Process conditions are developed for the manufacture of composite reinforcement for oak. Commingled glass and polypropylene fibers are consolidated in-situ on the surface of oak. Processing times from 30 to 120 seconds and pressures from .34 MPa to 1.4 MPa are tested. Micrographs of the composite and bond line reveal anisotropic fiber distribution in the composite, dry reinforcing fibers, voids, and incomplete consolidation. These microstructures are correlated with the processing problems which cause them, including insufficient heating time, poor matrix/glass mixing, and insufficient pressure to suppress void development. Lap shear strength and failure modes are related to microstructural features.

KEY WORDS

commingled fibers, Quercus, Oak, polypropylene, glass, thermoplastic, composite, reinforced lumber.

INTRODUCTION

Reinforcing skins can significantly improve compound lumber (1). The reinforcement supports joints in the lumber, preventing them from failing before the rest of the structure. The reinforcement also decreases the variability in failure loads of jointed compound lumber. Reinforced wood structures can be designed with smaller safety margins than can conventional lumber. The thickness of wood in a structure can therefore be reduced without sacrificing reliability. Reinforced lumber has great potential for use in weight critical applications. Several different reinforcing systems have been considered by different investigators, including metal rods, metal sheets, and carbon and glass in thermosetting polymer based composites (2,3). None of these reinforcement techniques has enjoyed economic success because of the high cost of materials and processing. This work develops a fast and economical process for manufacturing reinforced lumber from low cost thermoplastic composites and adhesives.

The processing of thermoplastic composites is difficult due to the viscous nature of the polymer matrix. Thermoplastics are difficult to impregnate into reinforcing fiber bundles. Impregnation difficulty can be overcome by using commingled glass and polymer fiber tows. When polymer fibers and reinforcing fibers are commingled in a single bundle, the distance the polymer has to travel in order to impregnate the reinforcing fibers is very small. Considerable work has been done to characterize the consolidation process for these materials (4-7). Of particular interest are studies on the formation and prevention of voids in the matrix (8,9). Voids form in the composite due to trapped air or outgassing (a separate phenomenon from dry reinforcing fiber bundles). Voids are suppressed when the pressure is sufficient to force the gasses to diffuse into the polymer.

This work develops a processing method by which commingled glass fiber/polypropylene tows are formed and bonded to oak in a single process. Commingled PP/glass yarn and an adhesive film are stacked on the wood surface with all the materials at room temperature. A heated press platen is then applied to the top surface of the commingled yarn. The pressure and heat from the platen melt the PP, consolidating the yarn and bonding the newly formed composite to the wood. Thermoplastic materials allow fast processing times compared to thermosetting materials. The consolidation and bonding process developed in this work can be completed in 3 minutes for a 2 mm thick skin. A similar reinforcement system made with thermosetting materials takes hours to cure.

MATERIAL SELECTION

Materials were chosen on the basis of the following three criteria: cost, processability, and availability. For a system to gain acceptance, it must be cost effective. For in-situ forming on the surface of oak, the polymer must be processible at temperatures below 200° C. Since the work is targeted toward applications, the raw material must be available in large quantities and have sufficient mechanical properties. The materials which best meets these requirements are polypropylene and glass fibers. Polypropylene is a widely available commodity thermoplastic. Glass fiber is a fiber of choice for low cost composite applications.

The PP/glass used for these experiments was Vetrotex Certainteed's Twintex® containing 60% glass by weight. The polypropylene's melting point is 163° C. The polypropylene's color was darkened with carbon black. For ease of handling in the lab, the material was collected in a unidirectional cloth with polyethylene cross stitching every 6 mm. The polyethylene stitching does not melt during processing. In some cross sections it is visible as inclusions in the composite. However, the failure surfaces of the lap shear samples do not show any indication that the stitching participates in the failure, and the best samples do not fail in the composite material at all.

The composite weight per area was held constant at 170 g of glass/ft^2. For well consolidated material, this weight translates to a 2 mm thickness of composite reinforcement. The thickness of the composite varied by ± 10 % according to void content and overflow of the molten PP. The consolidation time results will therefore be specific to the one weight tested. However, the observed morphologies are useful for developing processing conditions for any thickness.

The tie layer was provided by Bemis Corporation (CV202-17). The tie layer was a blend consisting predominantly of isotactic polypropylene and ethylene-propylene copolymer. The copolymer accounts for approximately 20% of the tie layer. The tie layer's melting point is 163° C. Polypropylene's non-polar chemistry has no natural affinity for wood (10). Pure PP in its molten state will form mechanical attachments to wood's surface irregularities but enjoys no chemical interaction. The ethylene-propylene copolymer does bond effectively with the wood surface. We introduce a thin film of the tie layer between the commingled fibers and wood surface. Because the flow properties and melting point of the film are very similar to polypropylene, the presence of the film does not add any difficulty in processing. The tie layer melts along with the PP fibers and bonds the composite and wood together.

The wood used for these experiments was oak (*Quercus*), both the red and white varieties. The two species are treated interchangeably in the experiments. The long grain was oriented along the fiber direction while the side grain varied randomly. The wood was allowed to acclimatize to ambient laboratory conditions. The wood imposes a temperature limit on the process; the surface chars at temperatures above 200° C.

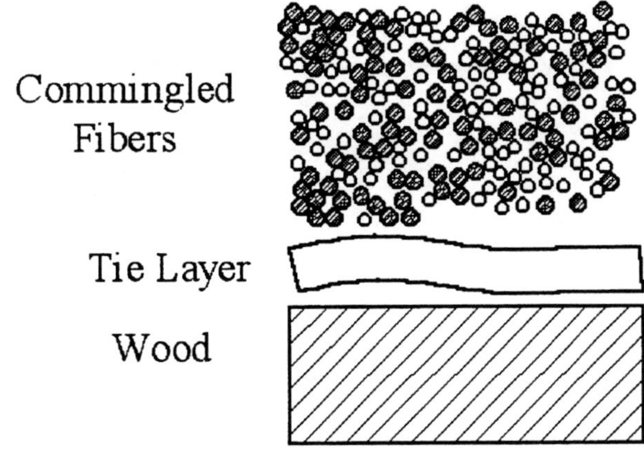

Figure 1: Raw materials configuration

PROCEDURE

The commingled fibers, tie layer, and wood were stacked together at room temperature as shown in Figure 1. The materials were then contact—heated from the top of the commingled fibers by a press and heating platen. The temperature of the heating platen was 240° C. Heat was conducted through the commingled fibers into the tie layer and then into the wood. Therefore the thermal history of any given point in the finished sample depended on its distance from the heating platen. Figure 2 shows temperatures during processing for the top and bottom (bonding) surfaces of the composite during a typical test. The bottom surface temperature lags behind the top surface temperature; thus, the lower portion of the composite has a shorter consolidation time than the upper portion. Processing times and temperatures were varied while cooling methodology, platen temperature, and material thickness were held constant.

Figure 3 is a schematic representation of the consolidation process and variables. Pressure was applied in two stages to minimize the time at high pressure. The platen was initially applied to the sample with a pressure of .17 MPa (25 psi) for 0, 30, 60, or 120 seconds. After the initial low pressure heating, the pressure was increased to .34 MPa, .69 MPa, 1.0 MPa, or 1.4 MPa (50, 100, 150, or 200 psi). The high pressure promotes good consolidation. A full factorial experiment was conducted with all the combinations of the above pressure and time variations. Six samples were prepared for each condition. One sample was used for microscopy, and the other five were tested in lap shear. Because of the difficulty in exactly repeating a pressure setting with our equipment, each group of tests at a given pressure was conducted as a block . Tests within each pressure block were conducted in random order of preheating and consolidation times.

After each sample was consolidated, a room temperature block of steel was placed on its surface to cool it. In less than 30 seconds the samples cooled enough for safe handling. The pressure during cooling ~7 kPa (~ 1 psi) was low enough to allow some warping of the composite due to thermal contraction. This warping caused voids to form at the bond line.

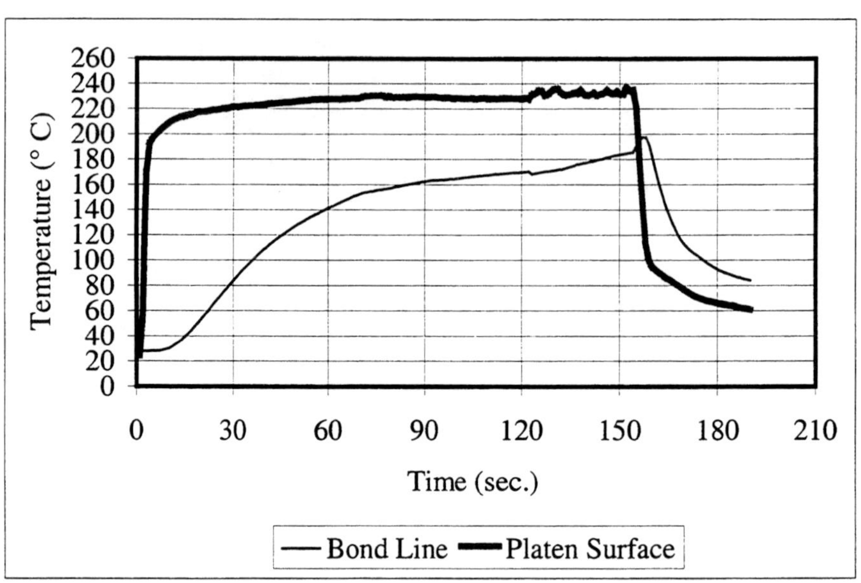

Figure 2: Typical temperatures during processing. The data here are from a sample heated for 120 seconds at 0.17 MPa, then for 30 seconds at 0.69 MPa, and then cooled for 30 seconds at 7 kPa.

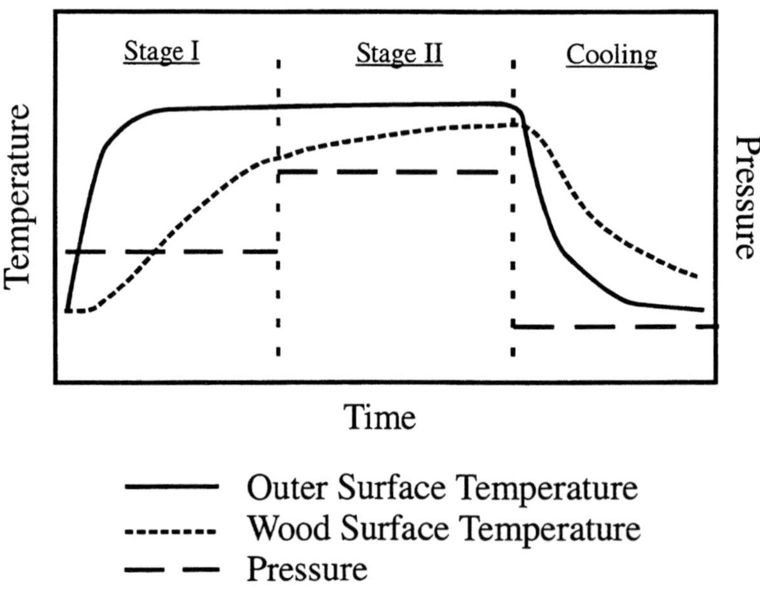

Figure 3: Schematic showing temperature trends of materials in contact with platen and in contact with wood during the three stages of the consolidation process.

In order to confirm that the voids observed at the bond line could be suppressed with sufficient cooling pressure, an extra set of 10 samples were prepared using high cooling pressure. The heating times were 120 seconds for first stage heating at .17 MPa (25 psi) and 30 seconds for second stage heating at .69 MPa (100 psi). The sample was then cooled with a room temperature block of steel applied with .69 MPa (100psi).

The test conditions in which the heating and consolidation stage-times total less than 90 seconds did not produce sufficient consolidation for the commingled fibers to bond to the wood. The melt front did not progress through the whole thickness of the fiber bed. For each condition which successfully bonded to the wood, six samples were prepared. Five samples were used for lap shear testing and the sixth was used for microscopy. Lap shear coupons were made with a 25 mm by 25 mm bond area, as shown in Figure 4. Samples were prepared for microscopy by cutting cross-sections perpendicular to the fiber and wood grain direction. Samples were encapsulated in epoxy for grinding and polishing. The cross–sections were then examined and photographed under a standard light microscope.

RESULTS

Figure 5 shows how the composite system evolves with time. The x–axis of the figure denotes increasing time and the y–axis denotes depth through the materials. A vertical line through the figure would capture the composite at a specific time during consolidation. The left side of the figure shows the beginning state of the system with commingled glass and PP fibers completely unconsolidated. The middle of the figure shows the composite mostly consolidated but with the tie layer not completely bonded to the wood. The right side of the figure shows complete melting of the matrix and good mating of the tie layer and wood surface. Typical problems with the composite and bonding are also shown, including anisotropic fiber distribution, incomplete consolidation, and voids in the body of the composite and along the bond line. Micrographs of the different morphologies sketched in figure 5 are shown in the figures which follow.

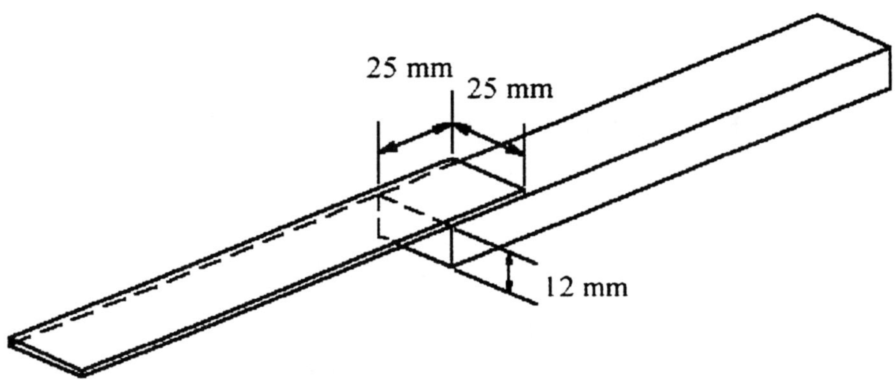

Figure 4: Lap shear sample configuration

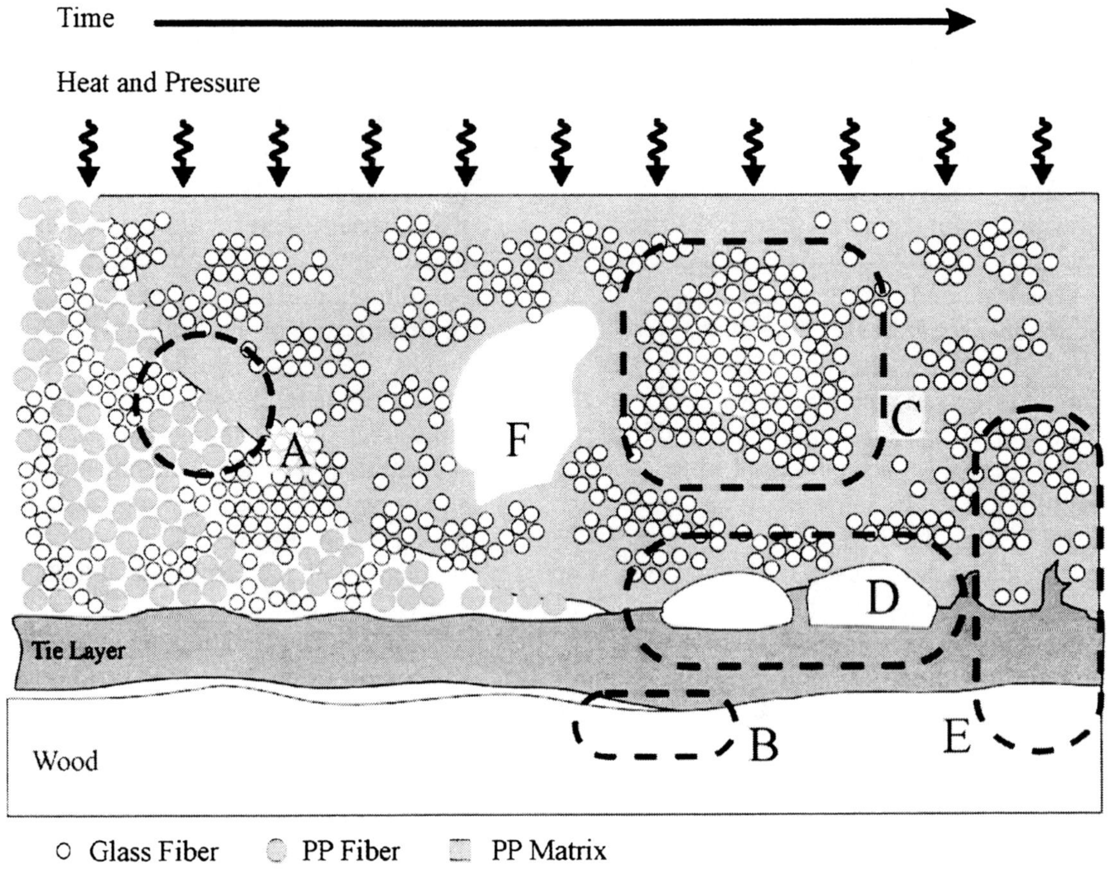

Figure 5: Schematic of evolution of composite microstructure showing various features including A) melt front, B) weakly bonded tie layer, C) dry fiber bundle, D) voids at the bond line, E) good consolidation, and E) voids in the bulk of the composite.

Consolidation of the commingled fibers and tie layer begins when the hot platen contacts the stacked materials. The force from the platen drives out air and compacts the fiber bed. The polypropylene fibers in contact with the hot surface melt. The melting fibers collapse and flow, allowing the fiber bed to compact further. The pressure and continued heating advance the melt front down through the commingled fibers and into glass fiber rich bundles. When the melt front reaches the tie layer, it too melts. Molten adhesive is pressed against the wood and soon invades any openings in the surface. When the hot platen is replaced with a room temperature platen, the system cools rapidly. The bond line generally cools to the melting point of the tie layer in less than 20 seconds. The composite tends to warp during cooling due to differential thermal contraction. This warping can pull the composite away from the surface of the wood, forming voids in the bond line.

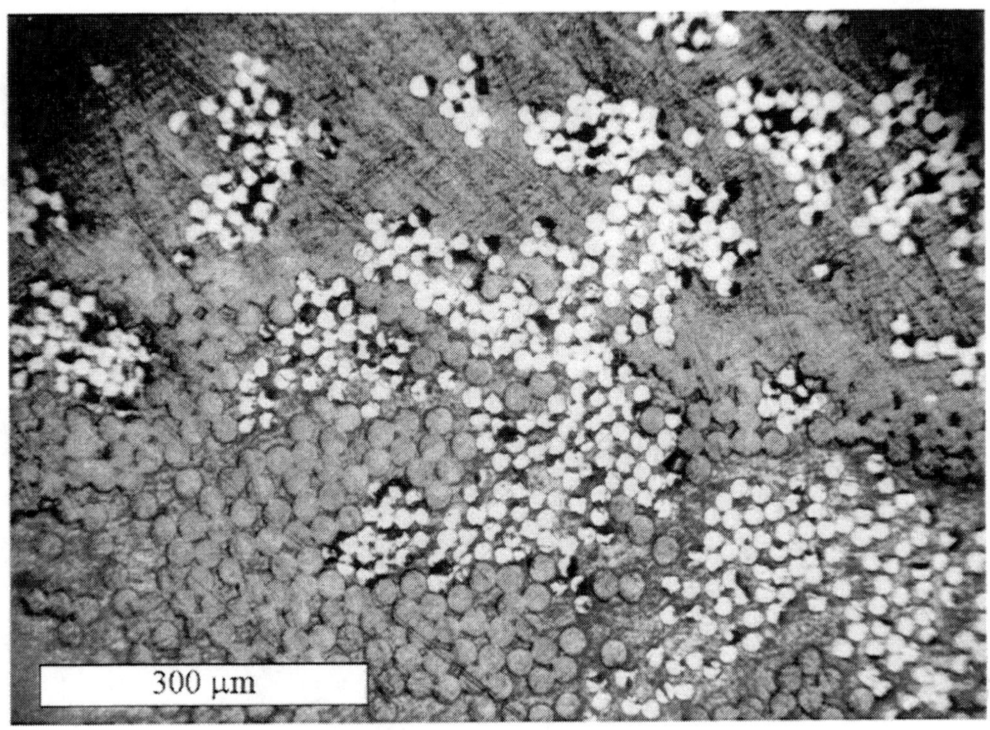

Figure 6: Consolidation melt front.

MICROGRAPHS

Figure 6 shows the melt front captured partway through the commingled fibers. This photo was taken from a sample heated for 30 seconds under .69MPa (100 psi). The melt front is advancing from the top of the photo toward the bottom. The upper part of the photo shows the consolidated composite while the lower part shows loose glass and PP fibers. The matrix around the loose fibers is the epoxy used to hold the sample for polishing.

Figure 7 shows a tie layer lifted away from a wood surface to which it was bonded. The sample started out with a bond between the wood and tie layer, but came apart when it was placed in epoxy to prepare it for grinding and polishing. There are unmelted PP fibers near the bond line and at the edge of the wood. There is no mixing of the tie layer and PP matrix. Figure 8 shows the failure surface of a lap shear coupon in which the tie layer did not fully wet the surface of the wood. Figure 8 is a view of the surface of the composite and tie layer that debonded from the wood, leaving a clean wood surface. The tie layer softened enough to be imprinted with the grain pattern of the wood. However, it did not bond well enough to tear as it pulled away from the wood. The melt front reached the tie layer but did not fully wet the wood and form a good bond. Insufficient wetting of the wood by the tie layer can be corrected by increasing the processing time.

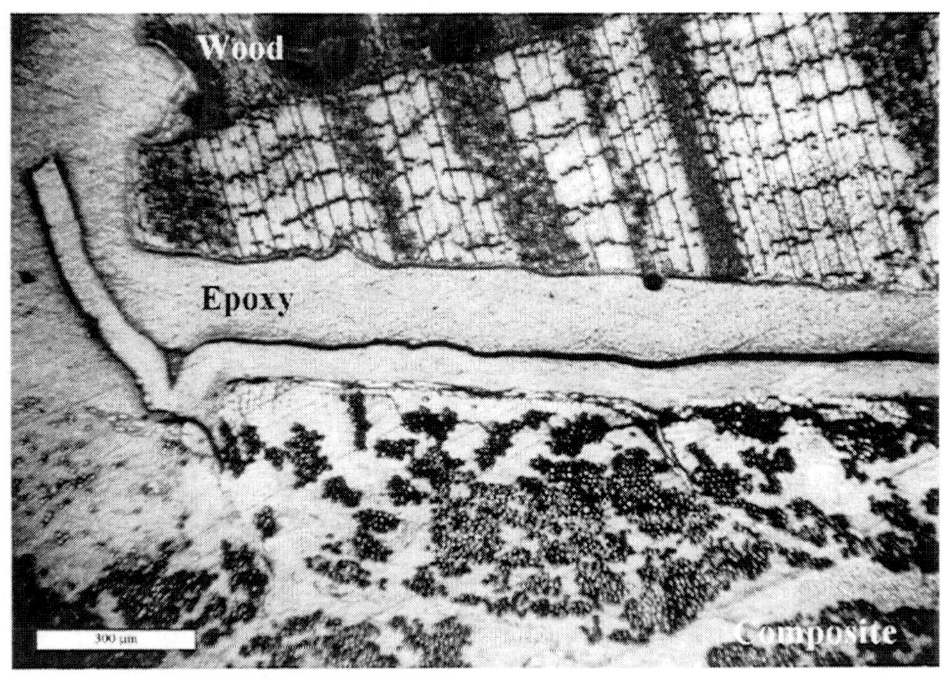

Figure 7: Insufficient wood-tie wetting.

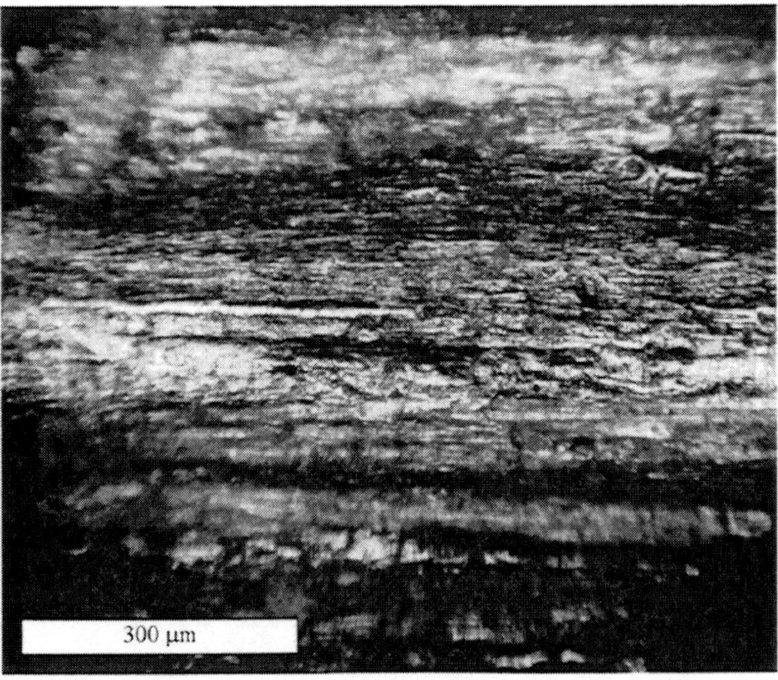

Figure 8: Tie layer surface debonded from wood

Figure 9 is an enlarged view of a dry fiber bundle. The PP matrix above the bundle and the tie layer below the bundle both show good consolidation, but there is no polymer among the clumped glass fibers. This is a common failing of thermoplastic composites made from commingled materials (4,5,7). Tens of minutes would be required to fully impregnate glass rich regions. As long as dry fibers are minimized, they do not cause problems, but if they are too extensive, then they provide an easy route for composite failure. Figure 10 shows a lap shear failure surface where the composite failed through a dry fiber bundle. This view of the wood surface shows dry glass fibers bonded to the wood by the tie layer and PP matrix. The fibers surfaces facing the fracture show no sign of PP adhesion.

Figure 11 shows extensive voids at the bond line. The voids show a subtle directionality, with the tendrils that bridge the gap between the wood and composite often having a slight tilt toward the centerline of the coupon. This photo was taken from a sample with first and second stage times of 120 and 60 seconds respectively. The pressure during the second stage was .69 MPa (100 psi). Voids were present at the bond line for all the different times and pressures except those cooled under high pressure. Figure 12 is a photograph of the failure surface of a lap shear coupon showing extensive bond line voids. The picture offers a downward view of the failure surface. The wave like texture of the polymer on the failure surface is created when the molten PP wets the wood surface and then pulls away during cooling. The opposing wood surface is covered with a thin layer of PP showing similar morphology.

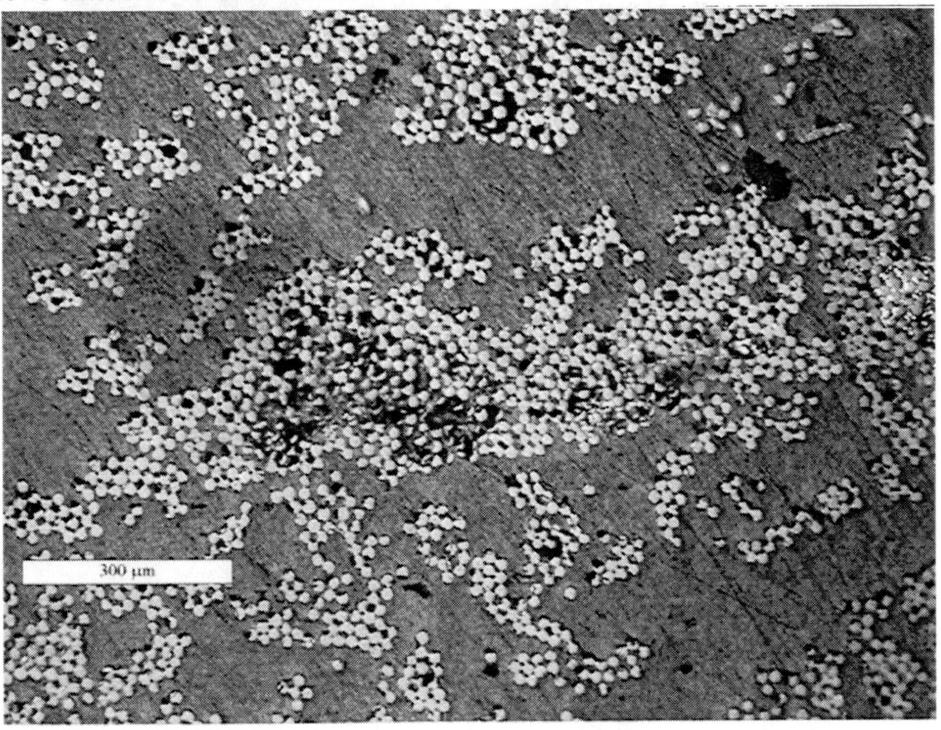

Figure 9: Dry fiber bundle.

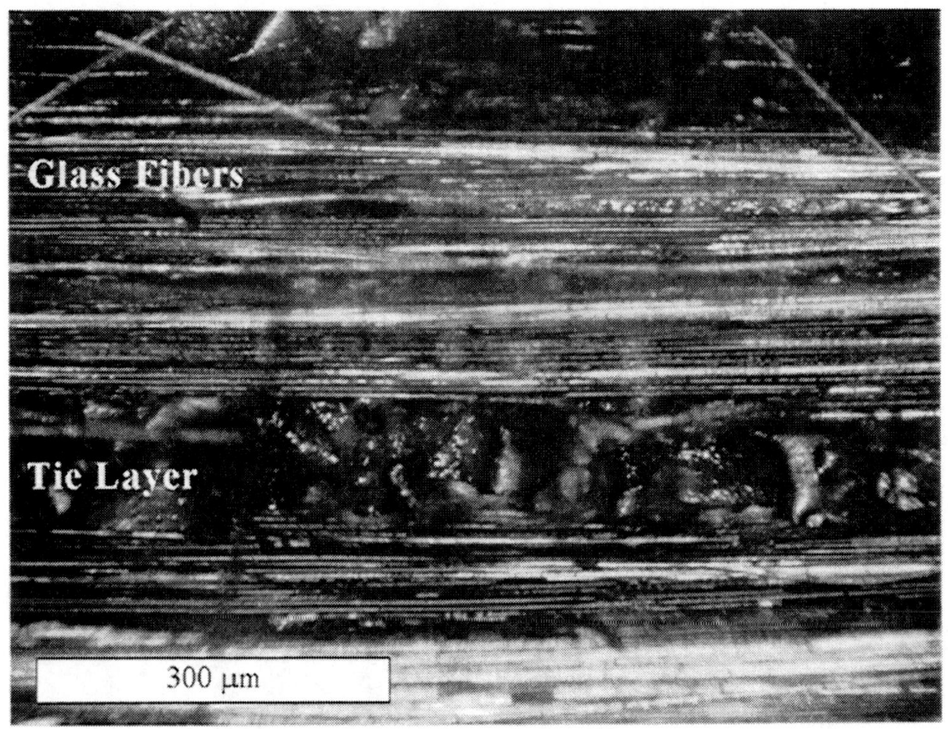

Figure 10: Lap shear failure through dry fibers

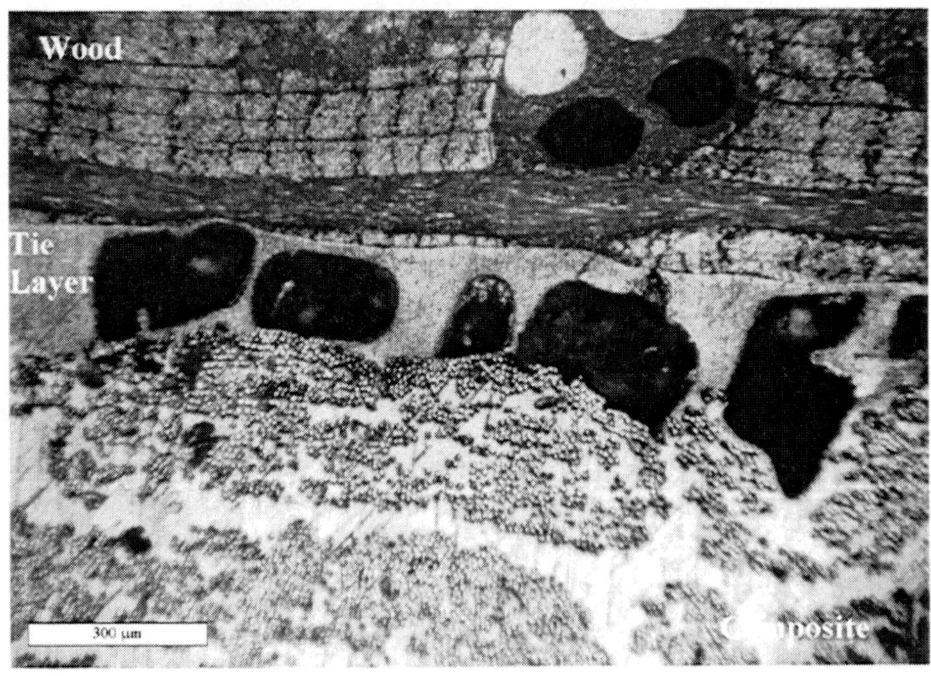

Figure 11: Voids at the bond line

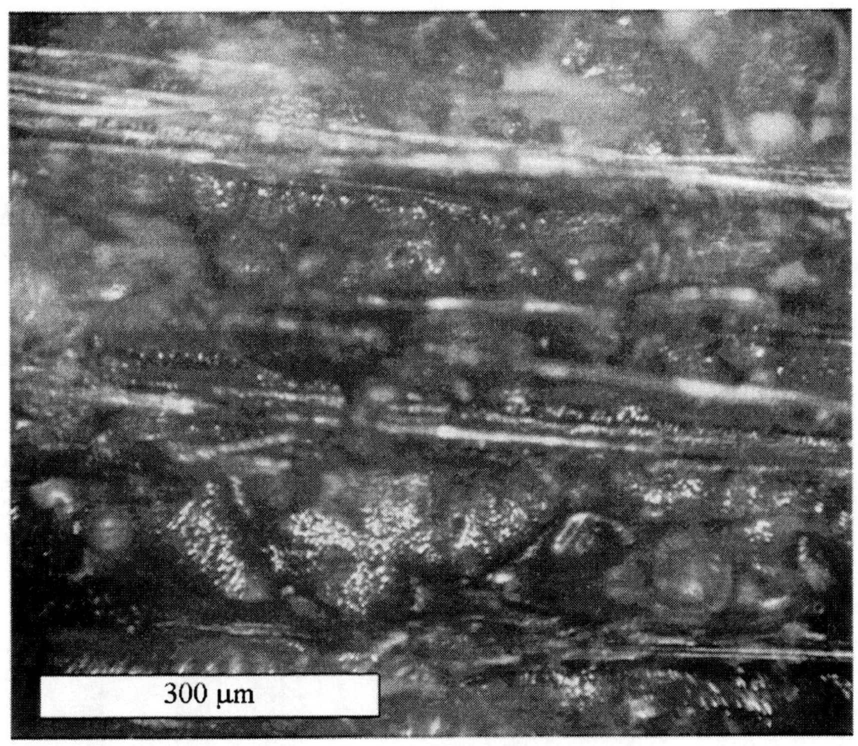

Figure 12: Lap shear failure through bond line voids.

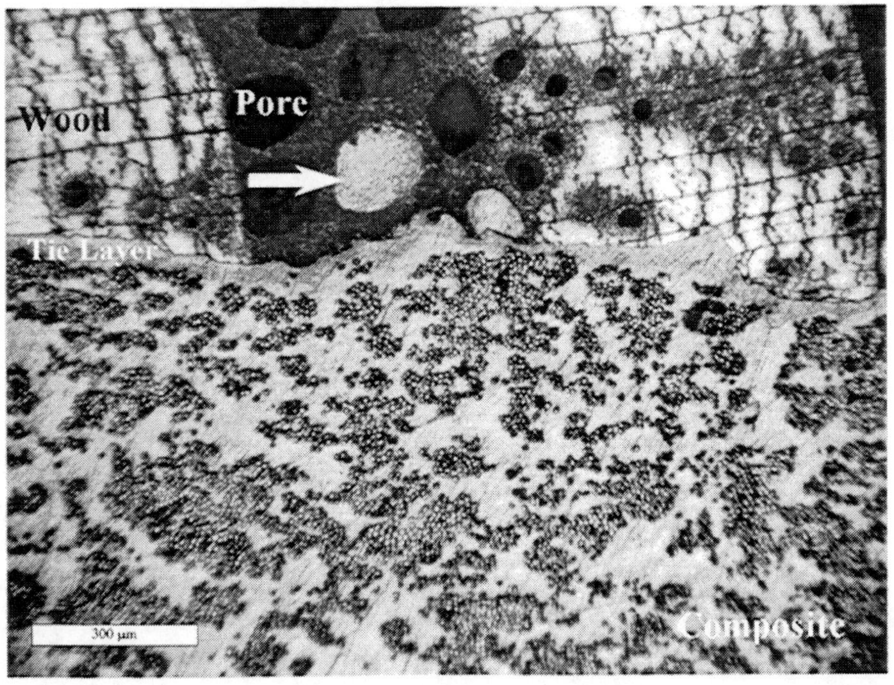

Figure 13: Good consolidation

Figure 13 shows the desired result– good consolidation and bonding. This sample was prepared using a consolidation stage pressure of .34 MPa (50 psi), first stage time of 120 seconds, and a second stage time of 30 seconds. Of note are the lack of voids in the composite or bond line. The tie layer shows some intermixing with the PP matrix and those glass fibers with which it has contact. In spite of the good consolidation, there are still regions of dry fiber. The horizontal arrow marks a large vessel in the wood that has been invaded by polypropylene. This sort of wood-polymer

strength results is an increase in strength as the sum of the heating and consolidation times increases. The proportion of processing time spent at low or high pressure does not seem to have an effect on the lap shear strength. With a single exception, differences in consolidation pressure have no effect either. Samples processed for 120 seconds with a high pressure of .34 MPa (50 psi) had an average strength 4.59 MPa (665 psi), significantly lower than 7.12 MPa (1030 psi), the average strength of all the samples processed for a total of 120 seconds. This may indicate that the lower pressure did not drive the

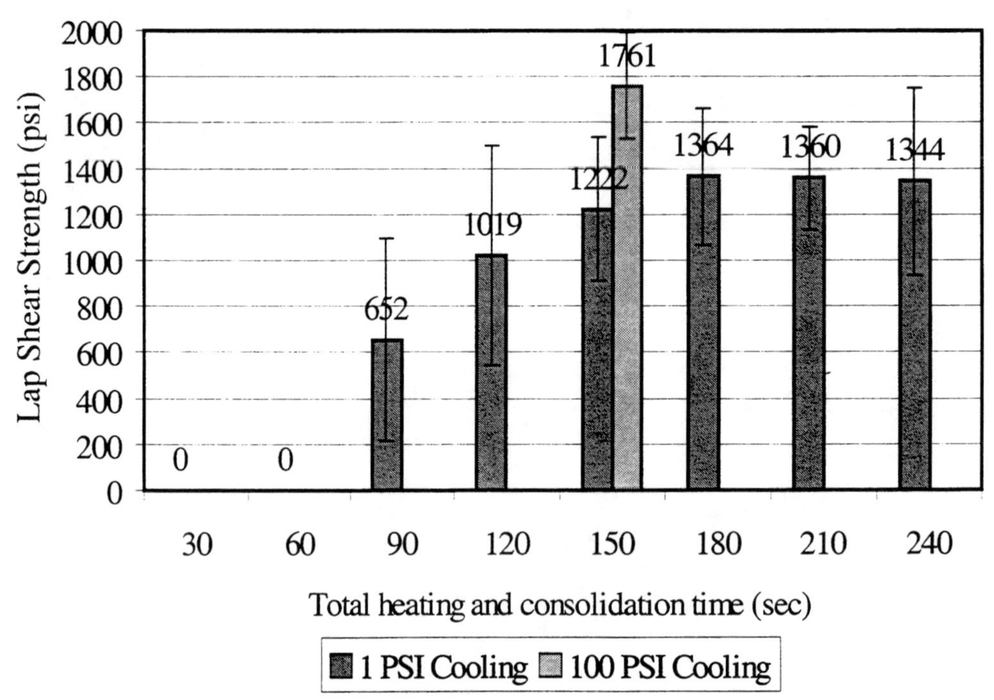

Figure 14 Lap Shear Strength

interaction will be explored further in the next chapter.

LAP SHEAR RESULTS

The lap shear strength values are tabulated in Appendix A. The only trend in the lap shear melt front as fast as the higher pressures. The effect of cooling pressure is significant. The set of samples processed with high pressure during cooling performed much better than the rest, having an average strength of 12.1 MPa (1760 psi). The lap shear samples showed a mix of failure modes.

Average strength values are shown in Figure 14. The results in the figure are averaged for each different total processing time. For example, the results for those samples with a first stage heating time of 60 seconds and a high pressure stage time of 60 seconds are averaged with those samples with zero seconds of low pressure heating and 120 seconds of high pressure heating. The number of samples averaged in each category is shown in Table 1. The numbers of samples are not all multiples of five because occasionally a set would have an extra sample or a sample would have to be

The lap shear samples displayed four distinct failure modes. These are cohesive failure of the wood, cohesive failure of the composite, cohesive failure of the tie layer, and debonding of the tie layer. Cohesive failure of the wood is the desired failure mode. If the system fails in the wood then there is no need for further improvement in the reinforcing skin and adhesive. Cohesive failure of the composite typically occurs through areas of the composite that are weakened by voids or extensive dry fibers. This type of failure indicates that the composite is insufficiently consolidated or that there are too many

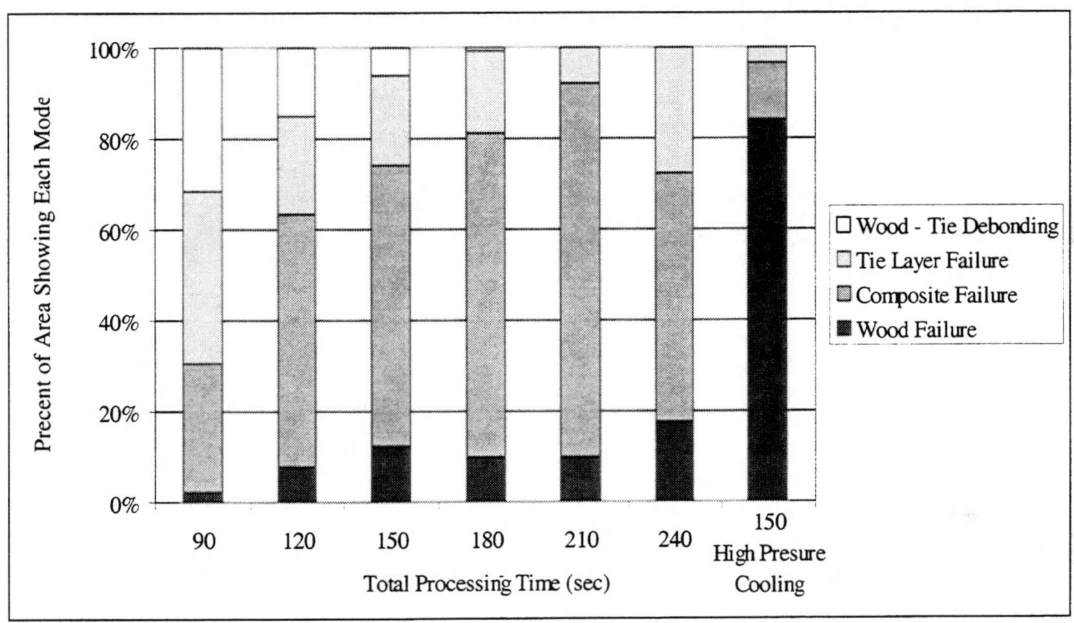

Figure 15 Lap Shear Failure Modes

discarded due to some processing defect. Samples which were processed for 60 seconds or less did not form bonds at all; the melt front did not reach the tie layer. The shear strengths of the samples which did bond show large scatter, probably due to the presence of voids at the bond line.

voids. Cohesive failure of the tie layer typically occurs through voids as shown in figure 11 and 12. De-bonding of the tie layer is due to insufficient interaction between the wood and tie layer as shown in figures 7 and 8.

The area of each failure mode on each sample was measured. Most samples had mixed failure modes with two or three types of failure mode represented in differing proportions. Figure 15 shows the proportions of each failure type, organized by processing time. The failure area proportions were averaged for sets of samples with the same total processing times.

The tie layer debonding failure mode tends toward decreasing occurrence with increasing processing time. Samples which were processed for a total time of 90 seconds failed do to tie layer debonding about 25% of the time. Samples processed for longer times showed progressively less failure due to debonding. The failure mode did not occur in samples processed for more than 180 seconds. This demonstrates that the debonding problem can be eliminated with sufficient processing time. The trend toward decreasing wood-tie debonding with increased processing time in figure 15 follows the strength trend in figure 14. The tie layer debonding occurs at a much lower stress level than cohesive failure of the composite, tie layer, or wood. The poorly bonded area in the samples therefore relates directly to their lap shear strengths.

The dominant failure mode for the samples processed for 180 seconds or more is cohesive failure of the composite. This failure mode indicates a weakness in the composite, usually due to voids. The last column in the chart shows the set of samples cooled with high pressure to prevent void formation. The failure mode shifts to predominantly cohesive wood failure with very little tie layer failure. The change in failure mode demonstrates that suppression of voids significantly increases the strength of the composite.

CONCLUSIONS

The two most significant flaws in the consolidation and bonding process were the insufficient wetting of the wood by the tie layer and the formation of voids near the bond line. These flaws are controlled by processing time and cooling pressure, respectively. The pressure during the second stage of consolidation within the experimental range had little if any effect on the strength of the composite and bond. The proportion of time spent at an initial pressure of .17 MPa (25 psi) versus the time spent at the high pressure during consolidation had little if any effect on the sample strength.

Insufficient consolidation time results in weak bonding of the tie layer to the wood. This results in premature failure of the bond line with little damage to the wood or tie layer. The problem can be corrected by increasing the consolidation time. For the 2 mm thickness of composite in this study the melt front reached the wood surface in 90 seconds, but more time was required for formation of a strong bond. A processing time of 120 seconds produced mixed results, but a processing time of 150 seconds produced consistently thorough consolidation and bonding.

Voids form near the bond line when the pressure during cooling is insufficient. The voids provide an easy route for fracture propagation through the tie layer and composite as seen in the failure surfaces. Voids contributed significantly to the noise in the lap shear strength results. These voids are the result of composite shrinkage during cooling. The outer surface of the composite cools first and begins to contract and warp. The change in dimension causes the composite to pull away from the surface of the wood. Cooling the samples under a pressure of .69 MPa (100 psi) eliminated the voids. The pressure was sufficient to hold the composite flat against the wood, overcoming the thermal stresses of cooling and the vapor pressure of any trapped air.

Dry reinforcing fibers were present in even the most well consolidated samples. However, for those samples which had few or no voids, the dominant failure mode was cohesive wood failure. Since the composite is not the weak link in the system, the dry fibers do not reduce the overall system capability. Therefore there is no need to increase the consolidation time simply to reduce the amount of dry fibers.

REFERENCES

1. W. Bulleit, *Wood and Fiber Science,* **16**, 3 (1984).

2. R. E. Rowlands, R. P. Van Deweghe, T. L. Laufenberg, and G. P. Krueger, *Wood and Fiber Science*, **18**, 1 (1986).

3. D. A. Tingley, and P. Eng, *International SAMPE Symposium & Exhibition*, **41**, 1 (1996).

4. F. N. Cogswell, and D. C. Leach, *SAMPE Journal,* May/June (1998).

5. V. Klinkmuller, M. K. Um, K. Friedrich, and B. S. Kim, *Proceedings of ICCM-10*, Whistler, B. C. Canada (1995).

6. V. Klinkmuller, M. K. Um, M. Steffens, K. Friedrich, and B. S. Kim, *Applied Composite Materials*, **1** (1995).

7. T. W. Kim, and E. J. Jun, *Advances in Polymer Technology*, **9**, 4 (1989).

8. Y. Leterrier, and C. G'Sell, *Polymer Composites*, **15**, 2 (1994).

9. A. Beehag, and Y. Lin, *Composites Part A*, **27A** (1996).

10. G. Han, and N. Shiraishi, *Mokuzai Gakkaishi*, **37**, 1 (1991).

APPLICATION OF HIGH FREQUENCY DIELECTRIC SPECTROSCOPY
FOR THE ASSESSMENT OF DURABILITY
OF ADHESIVELY BONDED COMPOSITE STRUCTURES

P. Boinard, R.A. Pethrick
Department of Pure and Applied Chemistry,
Thomas Graham Building,
University of Strathclyde,
295 Cathedral Street, Glasgow, G1 1XL, UK.

W.M. Banks
Department of Mechanical Engineering,
James Weir Building,
University of Strathclyde,
75 Montrose Street, Glasgow, G1 1XJ, UK.

ABSTRACT

Composites materials, especially adhesively bonded structures are being used in a wide range of applications, including situations where they become exposed to high levels of moisture. Ingress of moisture into a polymeric material will in general lead to changes in its mechanical properties usually associated with plasticisation effects. In bonded composite structures, water ingress can influence not only the mechanical properties of the matrix but also those of the matrix-fibre and adhesive-adherent interfaces.

Over the last ten years, the application of high-frequency dielectric techniques to the characterisation of the integrity and durability of adhesively bonded metallic structures has been extensively investigated by the co-authors.

In general, a bonded structure resembles a wave-guide in which the adhesive layer is the dielectric. Changes in the characteristics as a function of time of exposure to the environment can be used to monitor the ageing of such structures. This paper discusses the application of the principles to the study of carbon fibre reinforced plastics (CFRP) adhesively bonded composite structures. Carbon fibre is in general less conductive than aluminium material. However, it is sufficiently conductive to sustain the propagation of high-frequency dielectric signals. The effect of changes in the surface alignment and subsequent bulk orientation of carbon fibres on the dielectric propagation has been investigated. The ingress of moisture in the raw materials and in the joint structure is presented. The high-frequency time domain response (TDR) analysis allows the integrity of the structure to be explored and a good correlation is shown between TDR analysis and gravimetric results.

This study indicates that the success obtained in the application of high frequency dielectric measurements to adhesively bonded aluminium structures is also applicable to CFRP bonded structures. The dielectric study not only indicates a new way to assess the state of such a structure but is also producing new insights into the application of TDR measurement to non-isotropic materials.

1. INTRODUCTION

The integrity of power and signal cables and related structures is typically assessed by the use of time domain electrical measurement techniques. In the past fifteen years development of vector network analysers operating in the frequency range 300 kHz to 3 GHz has allowed accurate measurements of the electrical permittivity and loss in wave-guide structures.

During the past decade, the application of high frequency dielectric spectroscopy to the assessment of aluminium adhesively bonded structures has been investigated. It has been established that information on the quality and integrity of bonded structures can be obtained (Pethrick et al., 1995-1997a-c). Its potential as a non-destructive technique to monitor ageing and degradation of adhesive joint structures during exposure to a harsh hot and humid atmosphere has been demonstrated (Pethrick et al., 1997a). Time domain analysis allowed identification of defects present in the bond line (Pethrick et al., 1996-1997a and b), whereas frequency domain analysis allow the ingress of water in the bond line (Pethrick et al, 1997c).

Degradation of bonded structures occurs when significant amounts of moisture are absorbed by the adhesive. Once the water penetrates the bond line, the adhesive properties can be altered in a reversible manner by plasticisation or in an irreversible manner by hydrolysis or crack and craze formation (Comyn, 1991 and Shaw, 1993). Generally, degradation at the adhesive-adherent interface occurs by displacing the adhesive or by changing the adherent surface chemistry. The diffusion of water in the adhesive tends also to modify the mechanical failure mechanism of the joint from a cohesive failure in the adhesive layer to an adhesive failure at the adhesive-adherent interface (Matthews et al., 1991 and Kinloch, 1987).

High frequency dielectric spectroscopy techniques proved successful in the study of ageing of aluminium-epoxy resin-aluminium bonded structures since the conducting aluminium adherent generates a wave-guide structure for the propagation of electromagnetic waves. However, the aerospace industry increasingly uses adhesively bonded <u>composite materials</u> in aircraft primary and secondary structures. There is, therefore, a requirement for the development of non-destructive techniques to assess the integrity of these bonded composite structures. This paper explores the potential of high frequency dielectric spectroscopy as a non-destructive evaluation (NDE) method for adhesively bonded composite structures. This paper also addresses how electromagnetic wave propagation may occur in composite structures.

2. EXPERIMENTAL METHODS

2.1 Materials
The carbon fibre reinforced plastic (CFRP) adherents were manufactured from Hexcel Composites Ltd unidirectional carbon fibre pre-impregnated film, trade named 914C-TS(6K)-5-34%. The pre-impregnated film contained 66% by weight of high-tensile surface treated carbon fibres. The adhesive system used for bonding the CFRP plates was a 3M structural epoxy system trade named Scotch-Weld Brand AF-163-2U, consisting of a nylon woven fibre supported epoxy resin.

2.2 Manufacturing Process
Carbon fibre pre-preg layers were used to produce a series of different lay-up designs using a vacuum bag procedure in an autoclave. Curing was achieved at a temperature of 170°C under a pressure of 700 kPa. A post-cure at 190°C for 4 hours was performed. Joints were manufactured by joining two carbon fibre reinforced plastic (CFRP) plates with a lay-up of adhesive films. The assemblage, placed in an autoclave chamber, was cured using a vacuum bag system at a temperature of 125°C and a pressure of 200 kPa for 1 hour. The final system was a single lap joint made of 150 mm length, 50mm width and 1.8 mm thick CFRP plates bonded over a 10 mm wide overlap by a 1 mm thick adhesive.

2.3 Gravimetric Measurements
As soon as manufactured, all the samples were placed prior to their ageing in a dessicator at room temperature in order to avoid absorption of moisture. Gravimetric measurements during ageing were performed by removing the samples from a water bath held at a constant temperature of 60°C and rapidly blotted and weighed using an electronic balance (Mettler AJ100) with an accuracy of ± 0.1 mg. The times for the weighing experiments were assumed sufficiently short not to influence the values of the mass measured.

2.4 Dielectric Measurements
Dielectric measurements were carried out in reflection mode over the frequency range 300kHz to 3GHz using a Hewlett Packard 8753A Network analyser. The system was calibrated using three independent standards whose reflection coefficients are known over the frequency range of interest: short, open circuit and a matched load at 50 Ω. Details of the measurement technique and theory have been presented elsewhere (Pethrick et al., 1995 & 1997b). The ratio of reflected to incident waves ρ for such wave-guide lines is given by Equation 1:

$$\rho(t) = \rho_i(t)\delta(t) - \left[(1-\rho_i)^2 \rho_T \sum_0^N (\rho_i \rho_T)^{N-1} \delta(t - NT) \right] \quad (1)$$

where ρ_i and ρ_T are the reflection coefficients at the input and end of the joint respectively, t is the time, T is the total transit time, N is the number of transits in a time t and δ is the Kronecker delta. The reflection coefficients ρ_i and ρ_T are defined by:

$$\rho_i = \frac{Z_l - Z_0}{Z_l + Z_0} \quad \text{and} \quad \rho_T = \frac{Z_T - Z_l}{Z_T + Z_l}$$

where Z_0, Z_l and Z_T are the characteristic impedance of the system, the joint and the line termination respectively. The latter is equal to infinity for free end, therefore ρ_T equals 1. Time domain data were obtained by using the network analyser's inverse Fourier transformation over the frequency range.

3. RESULTS AND DISCUSSION

3.1 Effect of Orientation of the Carbon Fibre Layers on the TDR
When good contacts exist between the fibres of CFRP laminates, the composite adherent structure is able to transmit electric current polarising the adherent/adhesive interfaces. In the first attempt, joints were designed so that the unidirectional fibres were in the direction of the wave

propagation (Figure 1, top). The limitation of this design was that the water diffused perpendicularly to the mechanical testing direction, resulting in a limited understanding of the effects of water on the degradation of the mechanical properties of the bonded structure.

A design, which eliminates this restriction, was required. In the new design investigated, the unidirectional fibres were orientated in the same direction as the water diffusion and mechanical testing directions, therefore perpendicular to the direction of the electromagnetic wave transmission (Figure 1, bottom).

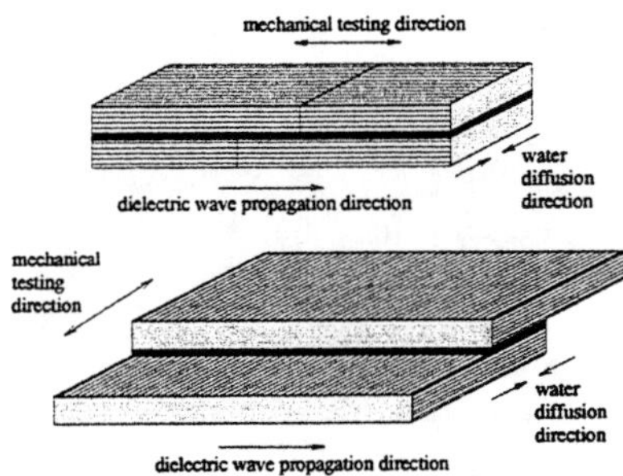

Figure 1: Previous joint design (top) and new joint design concept (bottom).

Results for unidirectional carbon fibre plate joints are presented in Figure 2 and several reflection peaks were observed (solid line). This is a typical TDR spectrum. When all the fibres are perpendicular to the dielectric wave propagation direction, the dissipation of the impulse energy in the first millimetres of the joint prevents the development of secondary reflection peaks. There is enough energy to percolate through the thickness of the plate, as shown by the presence of a first upward peak, but the high resistivity of the laminate matrix prevents propagation of the pulse along the joint.

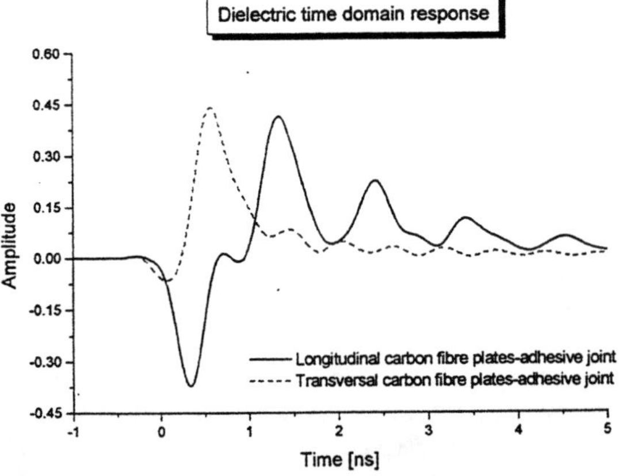

Figure 2: TDR of longitudinal and transversal CFRP joints.

The next design investigated consisted of improving the conductivity of the adherent in order to propagate the signal along the joint without significantly affecting the mechanical testing and involved orientation of the fibres in the propagation direction.

During high frequency dielectric spectroscopy, an electric field appears between the CFRP plates producing a magnetic field, which propagate a transverse electro-magnetic (TEM) wave at the adherent/adhesive interface. This part of the theory has not been previously explored since only isotropic materials such as aluminium have been studied. By using composite materials, such as carbon fibre pre-pregs, there is the possibility to modify the adherent/adhesive interface by changing the orientation of the fibre of the laminae at the interface.

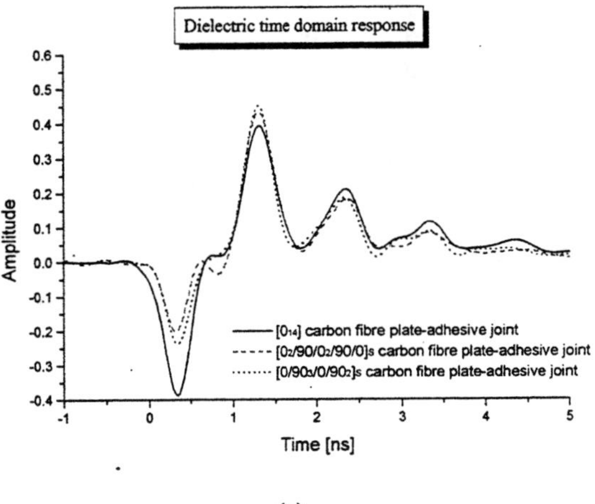

(a)

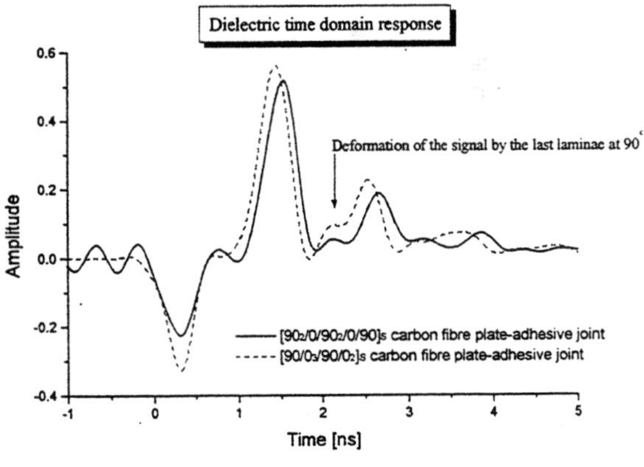

(b)

Figure 3: TDR of different carbon fibre lay-up: (a) laminae at the interface at 0° and (b) laminae at the interface at 90°.

For this purpose, five different lay-ups following the sequences $[0_{14}]$, $[0_2/90/0_2/90/0]_S$, $[0/90_3/0/90_2]_S$, $[90_2/0/90_2/0/90]_S$ and $[90/0_3/90/0_2]_S$ were manufactured. The differences between these lay-ups, which modify the overall dielectric property of the adherent, are designated in the direction of the last laminae at the interface and the percentage of fibres in each direction. The percentage of fibres in the 0°direction in each of the lay-ups is 100%, 74%, 26%, 26% and 74%, respectively. It should be noted that the directions of the carbon fibres are given relative to the dielectric wave propagation direction. A 90° fibre direction means that the fibres are perpendicular to the dielectric wave propagation (Figure 1, bottom). Figure 3a presents the TDR spectrum of the $[0_{14}]$, $[0_2/90/0_2/90/0]_S$ and $[0/90_3/0/90_2]_S$ lay-up carbon fibre plate-adhesive joint (respectively 100%, 74% and 26% of the carbon fibre at 0°) with the direction of the fibre of the laminae at the interface being at 0°. Figure 3b presents the TDR spectrum of the $[90_2/0/90_2/0/90]_S$ and $[90/0_3/90/0_2]_S$ lay-up carbon fibre plate-adhesive joint (respectively 74% and 26% of the carbon fibre at 0°) with the direction of the fibre of the laminae at the interface being 90°.

The sequence of the layers does not have an influence on the response and, as seen in Figure 3b, the presence of some layers at 90° does not significantly disturb the response when comparing the $[0_{14}]$ response with the other laminate responses. From these results, the $[90_2/0/90_2/0/90]_S$ configuration was selected for the adherent plates of the new joint design. This implies 14 layers of carbon fibres, 10 at 90° and 4 at 0°, representing 74% of fibres at 90° and 26% at 0°. Figure 4 presents a comparison between the TDR of the initial design and the new design.

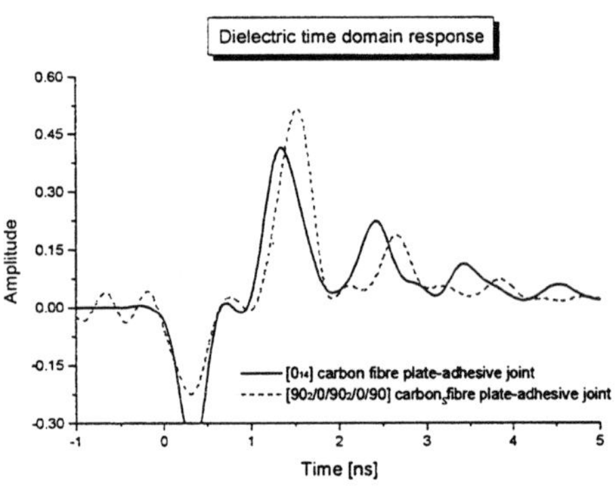

Figure 4: TDR of [0₁₄] and [90₂/0/90₂/0/90]s CFRP joints.

The lower amplitude in the impulse (downward) peak indicates a better balance between the impedance of the measuring system and the joint structure. The higher amplitude of the first response (upward) peak corresponds to an improved polarisation of the CFRP-adhesive interfaces. It is believed that the percolation between the fibres at 90° minimises the energy required for the electrical wave to reach the adherent-adhesive interfaces, consequently optimising the polarisation of the adhesive. The displacement of the peaks towards longer response time is consistent with the high resistivity of the CFRP matrix, which decreases the velocity of the TEM wave.

3.2.1 Carbon Fibre Layer Orientation Effect on Water Absorption, Figure 5 presents the results for different bulk orientation of carbon fibre layers in the composite adherent. The two designs presented are following the sequence $[0_2/90/0_2/90/0]_S$ and $[90_2/0/90_2/0/90]_S$. A significant difference between the plates is the amount of fibre ends exposed on each side of the plates. CFRP plates from the second design have more fibre ends in the length of the plates and therefore present a greater proportion of fibre ends to the ageing environment than plates from the first design. The $[90_2/0/90_2/0/90]_S$ plates would consequently be more sensitive to water absorption by capillary mechanism. As shown in Figure 5, the initial rate of water uptake is higher for the second design. This confirms the concern raised previously in section 3.1 regarding the effect of the amount of fibre ends exposed to the ageing medium. However, after long time exposure, the two designs present the same rate of absorption and the same amount of water absorbed. This indicates that the matrix is at an equilibrium value.

Figure 5 also shows that all the plates are following a non-Fickian behaviour. The weight of the plates does not reach any equilibrium but steadily increases. This behaviour has been previously observed in carbon fibre–epoxy matrix systems (Cai and Weitsman, 1994). The increase in weight may be generated by the formation of clusters inside the material, due to disbonding of the fibre/matrix interface.

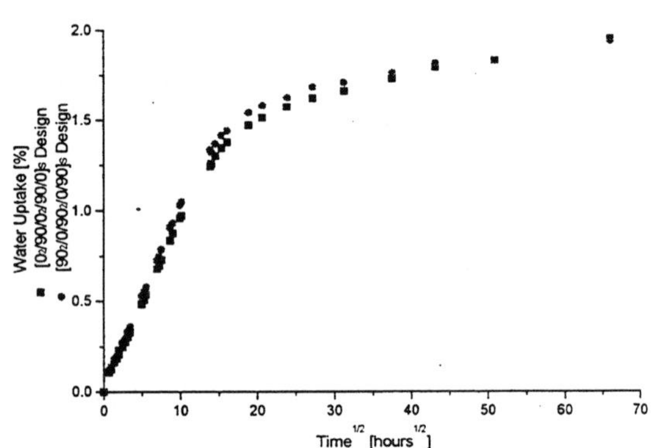

Figure 5: Water uptake in different CFRP plate designs, (■) [0₂/90/0₂/90/0]s and (●) [90₂/0/90₂/0/90]s.

3.2.2 Effect of Adhesive Post-Cure on Water Absorption, Figure 6 presents the water uptake profile for non-post-cured and post-cured adhesive. The large deviation around the value shown by the non-post-cured adhesive confirms the poor quality of the three-dimensional network of the epoxy resin, which results in a significant variation of the water absorption value. After a post-cure at 145°C for 8 hours, the water absorption of the adhesive was significantly decrease and more consistent results were obtained from sample to sample, as shown by the smaller deviation around the mean.

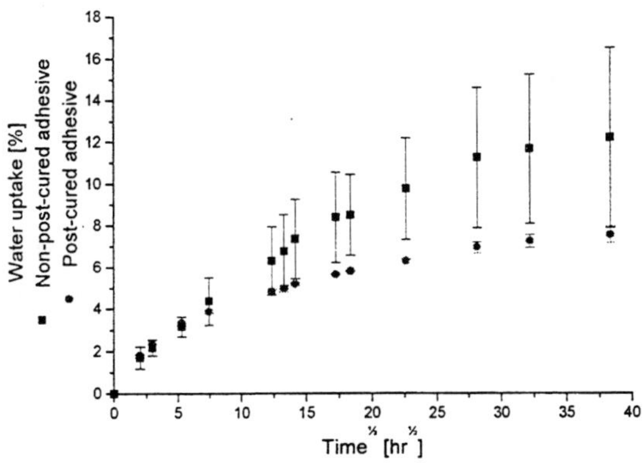

Figure 6: Water uptake in (■) non-post-cured and (●) post-cured adhesive.

3.2.3 Moisture Absorption in Joint Structure, Figure 7 presents the results for the joint structure, the CFRP adherents alone and the adhesive. The amount of water present in the adhesive has been estimated by subtracting the amount of water present in the joint to the amount of water present in the CFRP plates. The latter has been calculated from gravimetric results of CFRP plates exposed to identical ageing conditions. From these results, it is possible to dissociate the amount of water in the adherent from the amount of water absorbed by the adhesive.

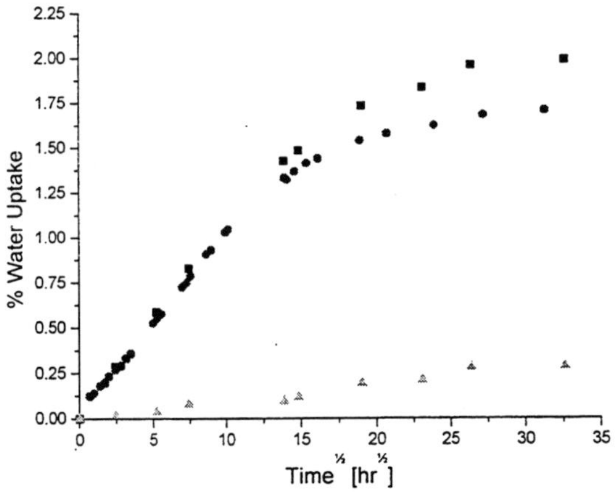

Figure 7: Water uptake of (■) the joint structure, (●) the CFRP adherents and (▲) the adhesive.

The adhesive in the joint absorbed less than the raw material. This is due to the difference in the absorption configuration: in the previous experiment all the surfaces of the adhesive sheet were in contact with the water whereas in the joint structure, only the edges are exposed to water.

3.3 Effect of Water on the TDR of Adhesively Bonded Composite Structures

TDR measurements were performed during ageing and are presented in Figure 8. The time difference between two neighbouring peaks is related to the effective permittivity $\bar{\varepsilon}$ of the material between the wave-guides (Pethrick et al., 1997a) by:

$$\bar{\varepsilon} = \left(\frac{c}{2 \cdot l / \Delta t} \right)^2 \qquad (2)$$

where c is the velocity of light in vacuum, l is the physical length of the joint, Δt is the time difference between the impulse peak and the first response peak and $\bar{\varepsilon}$ is the average dielectric permittivity. Figure 8a shows the displacement of the response peaks towards longer response time values. This is caused by the evolution of the dielectric permittivity of the adhesive due to the absorption of water which decreases the electrical wave velocity, consequently increasing the response time. Figure 8b presents the evolution of the average dielectric permittivity as a function of exposure time.

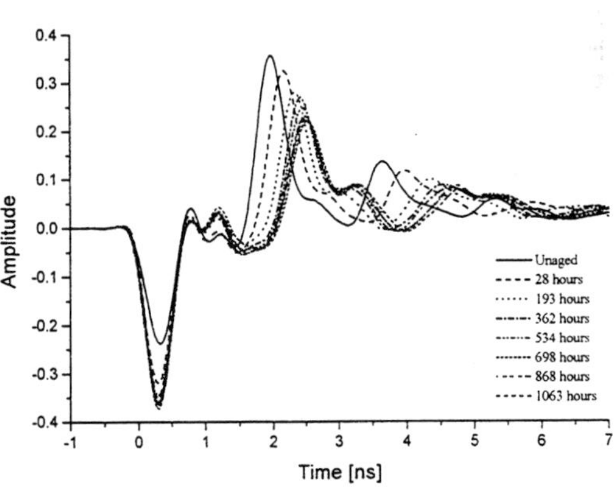

(a)

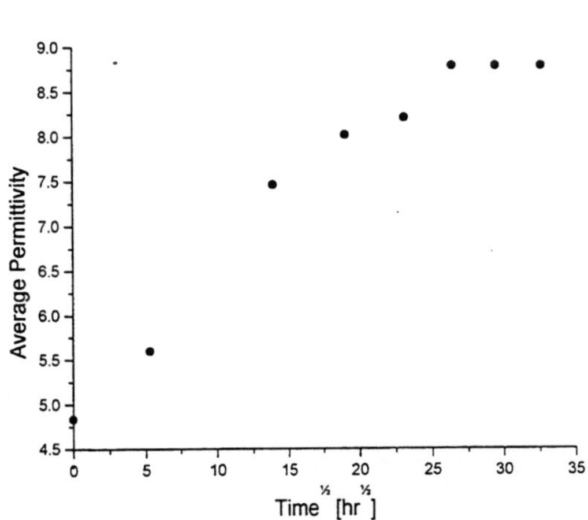

(b)

Figure8: Evolution of (a) the TDR and (b) the average permittivity during ageing.

3.4 Correlation Between Water Uptake and TDR of Adhesively Bonded Composite Structures

In order to compare the time domain results with the gravimetric results, normalisation of the data was performed using Equation 3:

$$Data_{Norm} = \frac{Data_t - Data_0}{Data_\infty - Data_0} \qquad (3)$$

where $Data_t$ is the value at time t, $Data_0$ is the initial value, $Data_\infty$ is the final value at equilibrium and $Data_{Norm}$ is the normalised value. A correlation coefficient of 0.96 between the TDR results with the gravimetric results shown in Figure 9 confirms that water penetration in adhesively bonded structure could be assessed by high frequency TDR measurement.

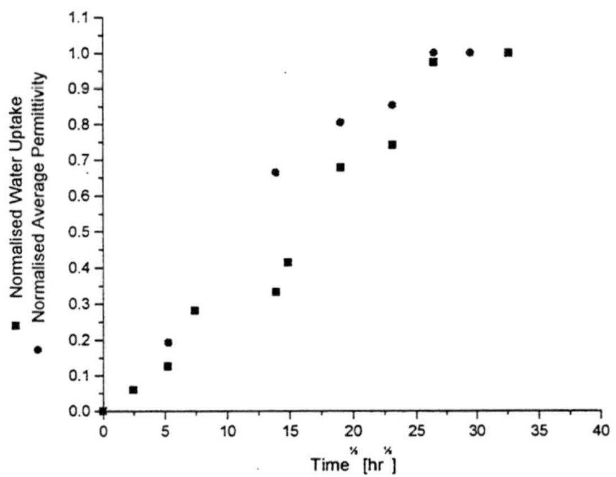

Figure 9: Correlation between normalised adhesive water uptake and average dielectric permittivity.

4. CONCLUSION

This study has shown the effect of the carbon fibre orientation on the high-frequency time domain analysis. It has been established that non-isotropic adherents, when well designed, provide good wave-guide for the TEM wave propagation. Correlation between the TDR and gravimetric data validated the use of TDR to identify the change in the dielectric permittivity of the bond line and to relate it to the water uptake in the structural adhesive.

This study on adhesively bonded composite structure has shown the potential of high frequency dielectric TDR as a non-destructive evaluation method to assess the integrity of the bond line present in aircraft primary and secondary structures.

5. ACKNOWLEDGEMENT

One of the authors (PB) wishes to thank the Non Destructive Evaluation Branch, Materials Directorate of the US Air Force for the provision of a maintenance grant in support of this study and British Aerospace for the provision of materials.

6. REFERENCES

Comyn J., 1981, 'The relationship Between Joint Durability and Water Diffusion', <u>Developments in Adhesives – Part 2</u>, Applied Science Publishers, London, pp. 279-314.

Kinloch A.J., 1987, <u>Adhesion and adhesives: Science and Technology</u>, Chapman and Hall, London.

John S.J., Kinloch A.J. and Matthews F.L., 1991, 'Measuring and Predicting the Durability of Carbon Fibre/Epoxy Composite Joints', <u>Composites</u>, Vol. 22, No. 2, pp. 121-126.

Banks W.M., Hayward D., Joshi S.B., Li Z-C., Jeffrey K. and Pethrick R.A., 1995, 'High-Frequency Dielectric Investigations of adhesive bonded structures', <u>Insight</u>, Vol. 37, No. 12, pp. 964-968.

Banks W.M., Dumoulin F., Hayward D., Li Z-C. and Pethrick R.A,1996, 'Non-Destructive Examination of Composite Joint Structures', <u>Journal of Physics Part D, Applied Physics</u>, Vol. 29, pp. 233-239.

Li Z-C., Hayward D., Gilmore R. and Pethrick R.A., 1997a, 'Investigation of Moisture Ingress into Adhesive Bonded Structures Using High-Frequency Dielectric Analysis', <u>Journal of Materials Science</u>, Vol. 32, pp. 879-886.

Li Z-C., Joshi S., Hayward D., Gilmore R. and Pethrick R.A., 1997b, 'High-Frequency electrical Measurements of Adhesive Bonded Structures – An Investigation of Model Parallel Plate Waveguide Structures', <u>Non-Destructive Testing & Evaluation International</u>, Vol. 30, No. 3, pp. 151-161.

Joshi S., Hayward D., Gilmore R., Yates L.W. and Pethrick R.A., 1997c, 'Environmental Ageing of Adhesively Bonded Joints: I. Dielectric Studies', <u>Journal of Adhesion</u>, Vol. 62, pp. 281-315.

Shaw S.J., 1993, 'Epoxy Resin Adhesive', <u>Chemistry and Technology of Epoxy Resins</u>, Blackie Academic & Professional, Glasgow, pp. 206-255.

Cai L-W. and Weitsman Y., 1994,' Non-Fickian Diffusion in Polymeric Composites', <u>Journal of Composite Materials</u>, Vol. 28, No. 2, , pp. 130-154.

Development and Characterisation of a Oxazolidone-Isocyanurate -Urethane Modified Diglycidyl Ether of Bisphenol A Epoxy Resin

K.S. Chian[*]*, S. Yi

School of Mechanical and Production Engineering
Nanyang Technological University
Nanyang Avenue
Singapore 639798
Republic of Singapore

Abstracts

Epoxy resins are very versatile polymers generally known not only for their adhesive and mechanical strengths but also their inherent thermo-oxidative stability. These properties have been widely exploited in many applications ranging from composites used aerospace components to electronics packaging. However neat epoxy materials suffer from relatively poor strain bearing ability and attempts to improve this brittle behaviour have been an area of many research interests. Toughening of epoxy using elastomers is not new and this is reflected in numerous publications and some have been successfully commercialised.

This paper presents a relatively new method of incorporating a rubber modifier into Diglycidyl ether of Bisphenol A (DEGBA) epoxy resin via the formation of an isocyanurate-oxazolidone-urethane structure. The synthesis involves several simultaneous reactions involving the following reactions: (a) epoxide and the isocyanate to form the oxazolidone, (b) trimerisation of the isocyanate to form isocyanurate and (c) isocyanate with polyols to form polyurethane. The synthesis and characterisation of the modified DEGBA resin will be presented. In addition, the effects of rubber content on the thermomechanical behaviour of the cured resin will also be discussed. In summary, it was found that the introduction of the isocyanurate-oxazolidone-urethane chemical groups into the epoxy molecular chain not only enhances the fracture toughness of the final resin but also improves on the thermal stability. This rubber-modified epoxy system shows excellent promise for adhesive and structural applications where thermal stability and fracture toughness are pre-requisites.

Introduction

Epoxy resins are thermosetting polymers used in a wide range of applications such as adhesives, coatings, laminates and moulding compounds, to name just few applications. These materials are known for their good adhesion to numerous substrates, high mechanical strengths, excellent chemical and electrical resistance. With increasing demands for greater performances from these materials such as thermal stability and improved flammability for applications such as the aerospace and electronics devices, attempts to improve these characteristics have often led to deterioration of other properties. As for example, increasing the crosslink density

---

[*] to whom all correspondence to be addressed.

of epoxy resins to achieve better thermal stability also lead to increased brittleness.

In spite of some of these known problems in epoxy resins, these polymers are still highly desirable because of their availability and well-established chemistry. In this paper we will present a relatively new approach to improve the toughness of epoxy-based thermoset by incorporating a polyurethane elastomer via oxazolidone and isocyanurate linkages. The effects of elastomer content on the final properties of the epoxy resin will be presented and discussed.

Epoxy resins (1-3)

The term epoxies are often given to compounds containing two or more highly strained three-membered ring structure involving oxygen and two carbon atoms, or the epoxide group. Figure (1) shows the molecular structure of a generic epoxy compound.

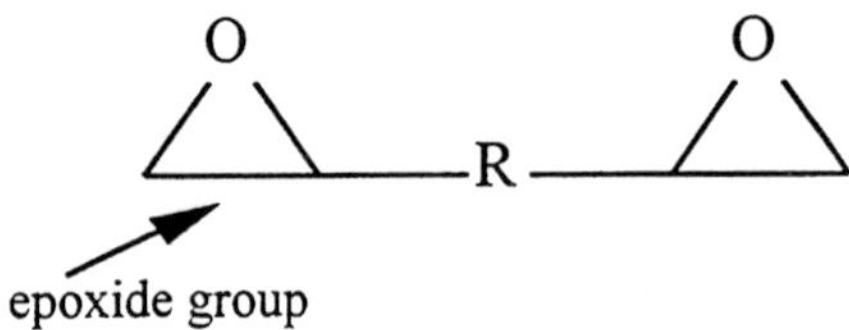

Figure (1): Molecular structure of a generic epoxy compound

Epoxy resins can be viscous liquids or brittle solids of low mechanical properties. The relatively high reactivities of the epoxide group render the ease of reactions with many curing agents to form three-dimensional networks. Basically the cured structure may be one of three types: (I) homopolymer whereby epoxide groups react mutually, (II) heteropolymer formed by the reaction between the epoxide and the curing agents or (III) a mixture of both reactions. Typical curing agents (also known as hardeners) used with epoxy resins include amines, acid anhydrides and their polymeric analogues. Catalysts are commonly used to accelerate the curing process and tertiary amines are widely employed.

Isocyanate Compounds

Isocyanate compounds are materials that contains the functional group –N=C=O. This functional group is known for their high reactivity with compounds containing active hydrogen such as water, polyols and amines. It can also react with the epoxide and anhydride groups to form oxazolidone and imide compounds.

Polyoxazolidone polymers (4-6)

These polymers are formed by the reaction between compounds containing the epoxide and the isocyanate group. Polyoxazolidones represent a new generation of thermally stable thermosets derived from isocyanate compounds. The presence of sterically hindered heterocyclic structures in the main backbone chain provide high stiffness and hence high thermo-mechanical properties. Figure (2) shows the chemical structure for an oxazolidone group.

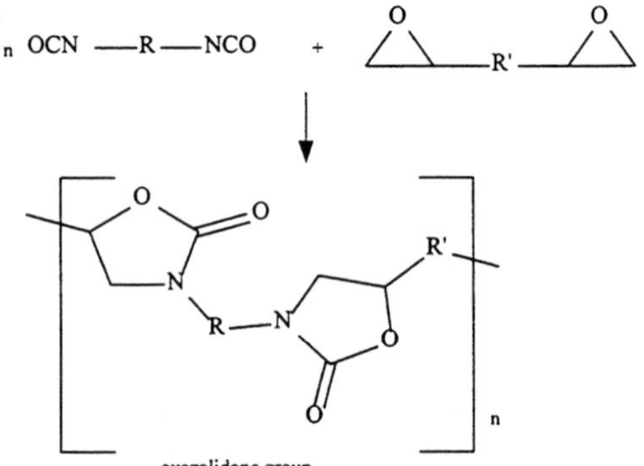

Figure (2): Cycloaddition of isocyanate to epoxide

Polyisocyanurate polymers (7,8)
Owing to the high reactivity of the isocyanate group, compounds containing these functional groups can, under certain conditions, undergo trimerisation reactions. Figure (3) shows the trimerisation reaction of an isocyanate group.

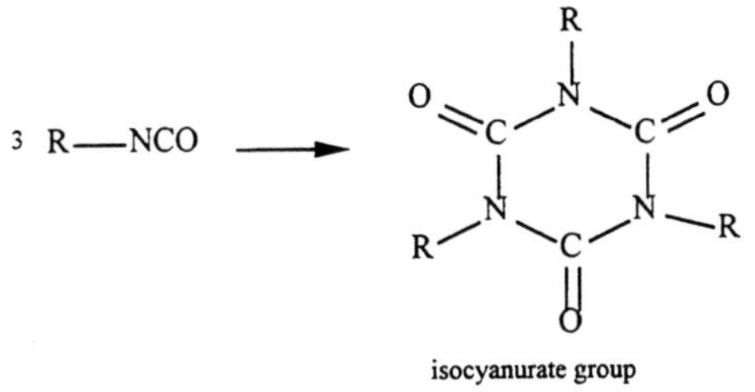

$$3\ R\text{---NCO} \longrightarrow \text{isocyanurate group}$$

isocyanurate group

Figure (3): Trimerisation reaction of the isocyanate group.

This six-membered ring trimer can lead to highly crosslinked structure that is thermally stable and the polymer is often considered to be flame resistant. Because of its inherently high crosslinked structure, polyisocyanurate polymers are extremely brittle.

Polyurethanes (7,8)
Polyurethanes are polymers containing the urethane functional group in their molecular backbones, i.e. materials containing the chemical structure – NHCO-. These polymers are known for their elastomeric properties and are mechanically very tough. The single most important characteristic that makes polyurethanes different from other polymers is their ease of synthesis. The properties of polyurethanes can be engineered through careful formulation of the reactants. These elastomers can either be thermoplastics or thermosetting depending on the functionality of the reactants used.

Commercially most polyurethanes are synthesised by the reaction between reactants containing polyisocyanate and polyols with the aid of catalyst. Figure (4) shows a typical synthesis of a polyurethane polymer.

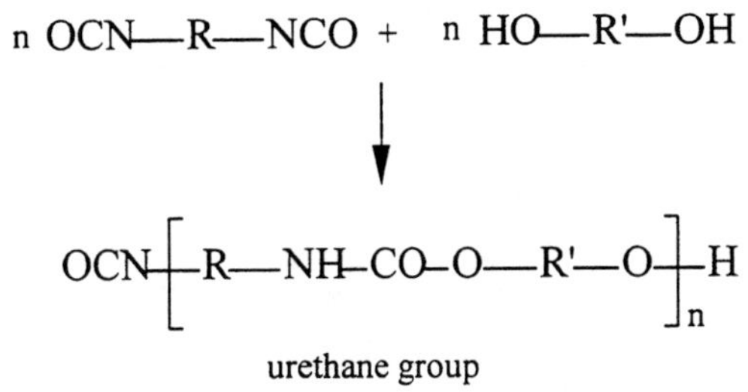

$$n\ OCN\text{---R---}NCO\ +\ n\ HO\text{---}R'\text{---}OH$$
$$\downarrow$$
$$OCN\!\!\left[R\text{---}NH\text{-}CO\text{-}O\text{---}R'\text{---}O\right]_{n}\!H$$

urethane group

Figure (4): Synthesis of polyurethane.

Objectives
The main objective of this paper is to report on the effect of incorporating a polyurethane elastomer phase into an isocyanurate-oxazolidone thermosetting polymer. The rubber-modified products will be characterised for their mechanical, thermal and water adsorption behaviours.

Methodology

Materials
Epoxy resin: Epomik R139, diglycidyl ether of bisphenol A (Mitsui Toatsu), equivalent weight 180-190.
Polymeric diphenyl methane diisocyanate (polyMDI): Voranate 288, isocyanate equivalent weight 131-136
Polyol: Polyether polyol, equivalent weight 345-355, hydroxyl value 160
Catalyst: Ankamine K54, a tris-2,4,6-dimethylaminomethylphenol (Anchor Chemicals)

All chemicals and glasswares used in the reactions are thoroughly dried in rotary evaporators and heated air oven respectively.

Samples Preparation

The amounts of polyurethane incorporated into the isocyanurate-oxazolidone (ISOX) system varies from 10 to 50 parts per hundred based on the weight of the ISOX used. Pre-determined amounts of the epoxy resin and polyol were carefully pre-mixed before being degassed in a vacuum oven at 80°C for approximately 30 minutes. The degassed mixture was then removed from the vacuum oven and sealed before allowing it to cool to room temperature in a dessicator. After cooling, 0.5wt % of Ancamine catalyst was added to the mixture and carefully stirred. Pre-determined amounts of polyMDI was then added to the epoxy-polyol-catalyst mixture and rapidly stirred to ensure good mixing. The reaction mixture was then poured into Teflon lined metal moulds and cure in a vacuum oven at 150°C for 15 hours. (The processing conditions were obtained from differential scanning calorimetry experiment. The temperature and time required to achieve a zero residual cure exotherm are used for curing all the samples.) After curing, the samples were removed and allowed to cool to room temperature in a dessicator where they are stored until needed for further tests.

Characterisation

All the samples were evaluated for their glass transition temperatures Tg, coefficient of thermal expansion CTE, thermal stability, and flexural properties.

- Tg of the samples were determined using a Perkin Elmer DMA 7. Sample was heated from -100°C to 250°C at a heating rate of 5°C/min. A dynamic loading frequency of 1 Hz and 10% strain was used.

- The coefficient of thermal expansion was determined using Perkin Elmer TMA 7 thermomechanical analyzer. Sample (5 x 5 x 3 mm) was heated from 0 to 200°C at a heating rate of 5°C/min in a quartz set-up. In the case of pure PU sample, starting temperature was -50°C. The load applied to the sample was set at 5mN.

- The thermal stability of the samples were evaluated on a TA Instrument TGA2950 thermogravimetric analyzer. The estimated life-time of the samples at 90°C, 150°C and 210°C in both air and nitrogen atmospheres were determined.

- Samples (80 x 8 x 3 mm) were evaluated for their flexural properties using an Instron 4206 universal tester. The span length was set at 45mm and the cross-head speed of 0.5mm/min was used throughout the test. The samples were tested until break.

Results

(a) Table (1) showed that the Tg of the PU-modified sample decreases with increasing amount of the PU phase.

(b) Table (2) shows that the thermal stability of the PU-modified samples decreases with increasing amount of the PU phase.

(c) Table (3) shows that the coefficient of thermal expansion of the sample increases with amount of PU-phase.

(d) Table (4) summarizes the mechanical properties of the various samples. It was found that as rubber content in the ISOX increases the moduli of the sample decreases but the fracture toughness increases.

Conclusion

It was found that the incorporation of the PU-phase has significant effect on the fracture toughness of the oxazolidone-isocyanurate modified DEGBA systems. It can be seen that at low levels of PU-rubber, say 10% PU, was effective in increasing the fracture toughness of the sample significantly in spite of reduction in other properties. In conclusion the preliminary investigation has shown that incorporation of PU is effective in improving the fracture toughness of an oxazolidone-isocyanurate modified DEGBA resin. Further works in improving the curing rate need to be investigated.

Acknowledgement

This work is supported by the NTU Applied Research Grant: RG 59/97.

References

1. B. Ellis (editor), "Chemistry and Technology of Epoxy Resins", Chapman and Hall, UK, 1994
2. Henry Lee and Kris Neville, "Handbook of Epoxy Resins", McGraw-Hill, 1982
3. Clayton A. May (editor), "Epoxy Resins Chemistry & Technology", 2nd edition, Marcel Dekker, NY, 1988
4. S.R. Sandler, F. Berg, G. Kitazawa, J. Applied Polymer Science, 9, 1994, (1965)
5. B.F. Speranza, W.J. Peppel, J. Org. Chem., 23, 1922, (1958)
6. J.S. Senger, I. Yilgor, J.E. McGrath, R.A. Patsiga, Am. Chem. Soc. Polym. Prepr., 26(1), 244, (1985)
7. C. Hepburn, "Polyurethane Elastomers", 2nd edition, Elsevier Applied Science, 1992
8. Gunter Oertel (editor), "Polyurethane Handbook", 2nd edition, Hanser, 1994

Table (1): Effect of PU concentration on the Tg of oxazolidone-isocyanurate-modified DGEBA.

Sample	Tg (°C)
Unmodified ISOX	158.53
PU(10 phr)-ISOX	154.57
PU(20 phr)-ISOX	138.36
PU(30 phr)-ISOX	131.60
PU(40 phr)-ISOX	120.27
PU(50 phr)-ISOX	113.28
PU	3.00

Table (2): Effect of PU concentration on the estimated life-time of oxazolidone-isocyanurate-modified DGEBA at elevated temperatures.

Sample	Estimated life-time (h)					
	Dry Nitrogen			In Air		
	90°C	150°C	210°C	90°C	150°C	210°C
Unmodified ISOX	6.0×10^8	2.5×10^5	676	4.2×10^8	1.9×10^5	595
PU(10phr)-ISOX	1.8×10^8	1.1×10^5	381	8.2×10^7	6.7×10^4	325
PU(20phr)-ISOX	1.2×10^8	8.9×10^4	393	4.1×10^7	4.1×10^4	231
PU(30phr)-ISOX	3.3×10^7	2.8×10^4	140	8.0×10^6	1.5×10^4	135
PU(40phr)-ISOX	2.4×10^7	2.2×10^4	116	3.0×10^6	8.7×10^3	108
PU(50phr)-ISOX	1.2×10^7	2.0×10^4	125	2.2×10^6	7.2×10^3	96
PU	5.9×10^5	1.6×10^3	20	5.7×10^3	112	6

Table (3): Effect of PU concentration on the CTE of oxazolidone-isocyanurate-modified DGEBA.

Sample	CTE (/°C)	
	Below Tg	Above Tg
Unmodified ISOX	9.55×10^{-6}	1.19×10^{-4}
PU(10phr)-ISOX	2.45×10^{-5}	1.64×10^{-4}
PU(20phr)-ISOX	4.87×10^{-5}	2.17×10^{-4}
PU(30phr)-ISOX	6.44×10^{-5}	2.72×10^{-4}
PU(40phr)-ISOX	7.12×10^{-5}	2.88×10^{-4}
PU(50phr)-ISOX	8.88×10^{-5}	2.95×10^{-4}
PU	-	1.38×10^{-4}

Table (4): Effect of PU concentration on the mechanical properties of oxazolidone-isocyanurate-modified DGEBA.

Sample	Modulus (MPa)		Yield Stress (MPa)		Yield Strain (%)		Toughness (MPa)	
	Mean	%Δ	Mean	%Δ	Mean	%Δ	Mean	%Δ
Unmodified ISOX	3127	-	122	-	4.1	-	0.29	-
PU(10phr)-ISOX	2530	-19.2	111	-8.9	4.8	19.3	0.33	14.1
PU(20phr)-ISOX	2047	-34.5	107	-12	5.9	46.7	0.41	39.5
PU(30phr)-ISOX	1836	-41.3	98.4	-19.2	6.6	62.5	0.44	50
PU(40phr)-ISOX	1675	-46.4	90.3	-25.8	7.11	75.6	0.45	54.5
PU(50phr)-ISOX	1460	-53.3	76.2	-37.4	8.44	108.4	0.49	67.2

MD-Vol. 88, Polymeric Systems
ASME 1999

Design and Analysis Techniques for Filament-Wound Composite Pressure Vessel Domes

William E. Howard
Graduate Student
Mechanical and Industrial
Engineering Department
Marquette University
Milwaukee, WI 53233

G.E.O. Widera
Professor
Mechanical and Industrial
Engineering Department
Marquette University
Milwaukee, WI 53233

ABSTRACT

The use of filament-wound composite pressure vessels has expanded into many new markets in recent years, creating the need for better design and analysis techniques, particularly for the end domes. In this paper, design and analysis techniques are developed for elliptical-conical dome profiles with planar filament winding patterns. The effects of wide winding bandwidth are included by dividing the band into sub-bands and considering any point on the dome contour to be a laminate made up of the sub-bands. The slippage tendency of the band at its edges is also calculated.

NOMENCLATURE

a semi-major axis of elliptical profile
A_{cyl} cross-section area of composite in cylinder
b semi-minor axis of elliptical profile
BW band width of composite winding band
d depth of dome
h_{cone} height of conical region of dome
N_ϕ stress resultant in meridian direction
N_θ stress resultant in hoop direction
r radial position of a point on the dome
r_o radius of polar opening
r_p radius to center of band near polar opening
r_1, r_2 radii of dome contour
s distance along fiber path
t composite thickness
t_{cyl} composite thickness in cylinder
u,v parameters used in calculation of α and t
x axial position of a point on the dome, measured from tangent
x_{cone} axial position to ellipse-dome interface
α orientation angle of fibers
α_c angle of fibers in cylinder
ϕ angle between axis of vessel and normal to dome surface
ϕ_{cone} ϕ at ellipse-cone junction

λ friction coefficient required to prevent slippage
μ friction coefficient
θ rotation angle of liner during winding

INTRODUCTION

Filament-wound composite pressure vessels have been used in many demanding applications such as rocket motor cases and oxygen tanks for scuba divers and fire fighters since the early 1960's. In the 1990's, a number of new markets have emerged for all-composite or composite overwrapped tanks, including reservoir tanks for home water systems and air tanks for truck brakes. Potentially huge markets such as compressed natural gas tanks for automobiles and hydraulic accumulators exist for this type of pressure vessel. Reliable methods for design and analysis need to be established in order for these new markets to be developed fully. A review of literature has shown that several aspects of composite vessel design have not been explored to any great extent. For example, effects of the width of the winding band on the stability of the winding pattern and on the stresses developed in the dome have not been studied. Also, most analyses of dome shapes focus on isotensoid shapes rather than simple shapes such as the ellipse-cone geometry proposed in this analysis.

The goal of this project is to present a design methodology which can be easily implemented by a designer who is familiar with the filament winding process. The design tools presented are limited to simple (ellipse-cone) design contours and include the effects on pattern stability and membrane stresses caused by the use of wide winding bands.

PRIOR WORK

Much prior research on filament wound domes has focused on isotensoid contours (Fukunaga and Uemura, 1983, and Hojjati et al, 1995). This contour is such that stresses in the fiber directions are equal anywhere on the dome contour. The stresses are based on netting analysis, in which contributions of the matrix material of the composite and the stiffness of any liner are ignored. The governing equations are solved numerically to define the dome contour point by point. The idea of aligning the fibers with the direction of the stress resultant seems to be logical, but several limitations prevent the perfect implementation of this scheme:

- There exists an inflection point along the dome contour, beyond which the theory breaks down.
- The matrix does contribute some amount of stiffness.
- Most vessels used in non-aerospace applications contain some type of isotropic liner, the stiffness of which is not insignificant. (Rocket motor cases usually have rubber insulating liners, the stiffness of which may be ignored.)
- The contour must be optimized based on a defined winding pattern. Therefore, any minor changes in the pattern definition will result in the contour being non-optimum.
- The optimum contour applies only to the centerline of the winding band. The orientation angles are different at the edges of the winding band. Most commercial applications utilize winding bands as wide as possible in order to maximize the amount of material placed during each pass and therefore reduce manufacturing time and cost.

Other composite dome contours that have been studied include spherical (Mistry and Levy-Neto, 1994) and elliptical (Logan and Hourani, 1983) contours. However, the referenced works do not address filament wound constructions but rather layups with constant thickness and fiber angles relative to position on the dome.

In all of the literature cited, calculations of orientation angle, thickness, slippage, and stresses are based on the centerline of the winding path. No consideration of how these properties are affected by a wide winding band were found.

This analysis will be restricted to elliptical dome contours. Since most filament-wound domes have polar openings and isotropic polar bosses to facilitate attachments, a conical region is included near the polar opening. This allows the definition of the polar boss geometry to be kept simple. An example of an elliptical dome with a conical

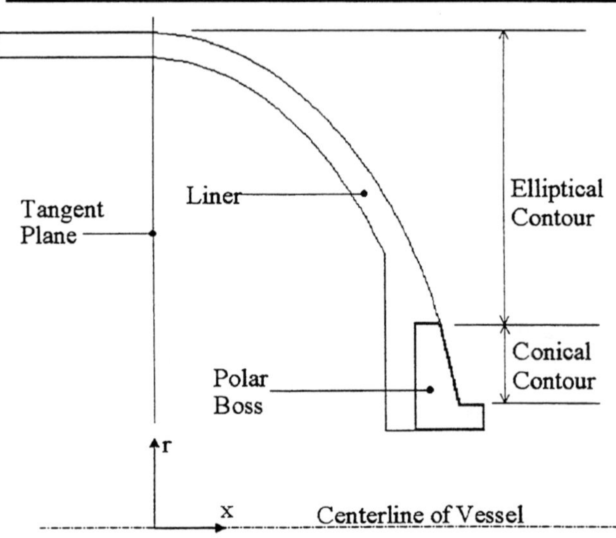

FIGURE 1 ELLIPTICAL-CONICAL DOME CONTOUR

section near the opening is shown as Figure 1. The distance from the cylinder-to-dome interface (the tangent plane) is defined as the x-coordinate.

The equations for orientation angle, thickness, fiber slippage, and stresses will be modified to consider the effects of bandwidth.

DOME GEOMETRY CALCULATIONS

The geometry of the filament-wound dome depends on the contour of the liner and the winding bands. Equations will be developed for the elliptical-conical contour and planar winding patterns in this section.

Elliptical-Conical Dome Contour

The ellipse is defined by the ratio of its semi-major (a) to semi-minor (b) axes. Therefore, a ratio of 1 is a sphere, and higher a/b ratios correspond to "flatter" dome contours. The definition of the elliptical contour is:

$$\left(\frac{x}{b}\right)^2 + \left(\frac{r}{a}\right)^2 = 1 \tag{1}$$

Solving for r as a function of x:

$$r = \frac{a}{b}\sqrt{b^2 - x^2} \tag{2}$$

The slope of the contour at any point can be found by differentiating Eq. (2):

$$r' = -\frac{ax}{b\sqrt{b^2 - x^2}} \tag{3}$$

where ( )' denotes differentiation with respect to x. Near the polar opening, a conical section is typically used. The use of the conical section allows for simplicity of the polar boss definition, and allows the use of a common polar boss for different sizes of vessels. The radial position of the junction of the elliptical and conical regions is defined as the sum of the composite opening radius r_o and the height of the conical section h_{cone}, as shown in Figure 2. Note that h_{cone} can extend beyond the top of the polar boss and into the liner if desired.

The angle ϕ defines the direction of the normal to the liner surface, as shown in Figure 2. From Eq. (3), the value of ϕ is:

$$\phi = \tan^{-1}\left(\frac{b\sqrt{b^2 - x^2}}{ax}\right) \tag{4}$$

Since the contour should transition smoothly from the elliptical region to the conical region, the cone angle ϕ_{cone} must equal the angle ϕ at the junction. Therefore,

$$\phi_{cone} = \tan^{-1}\left(\frac{b\sqrt{b^2 - x_{cone}^2}}{ax_{cone}}\right) \tag{5}$$

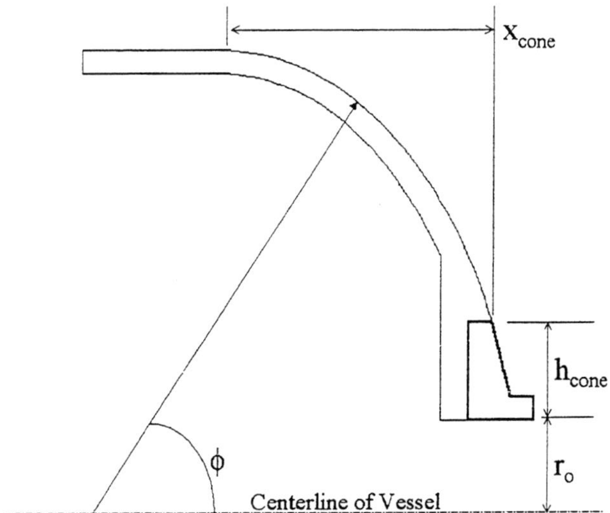

FIGURE 2 ELLIPSE-TO-CONE JUNCTION LOCATION

From the definition of the elliptical contour given in Eq. (1),

$$x_{cone} = \frac{b}{a}\sqrt{a^2 - \left(r_o + h_{cone}\right)^2} \qquad (6)$$

Once the cone angle ϕ_{cone} has been established, the radial coordinate r in terms of x in the conical region can be written as:

$$r = r_o + h_{cone} - \frac{x - x_{cone}}{\tan\phi_{cone}} \qquad (7)$$

Differentiating Eq. 7 with respect to x yields:

$$r' = -\frac{1}{\tan\phi_{cone}} \qquad (8)$$

Planar Winding Path

A geodesic winding pattern may be defined by applying Claiaut's equation (Lossie and Van Brussel, 1994):

$$r\sin\alpha = \text{constant} \qquad (9)$$

where α is the orientation angle of the fiber relative to the meridian of the dome (the winding angle). One of the problems with defining a geodesic pattern is that it must be defined to the winding machine software point by point. This point-by-point definition results in uneven accelerations of the winding machine motions, resulting in overall slower winding speeds. Also, the no-slip condition applies only along the geodesic path (usually defined as the centerline of the winding band). The stability of the winding band still depends on the coefficient of friction at the edges of the band. Also, many vessels

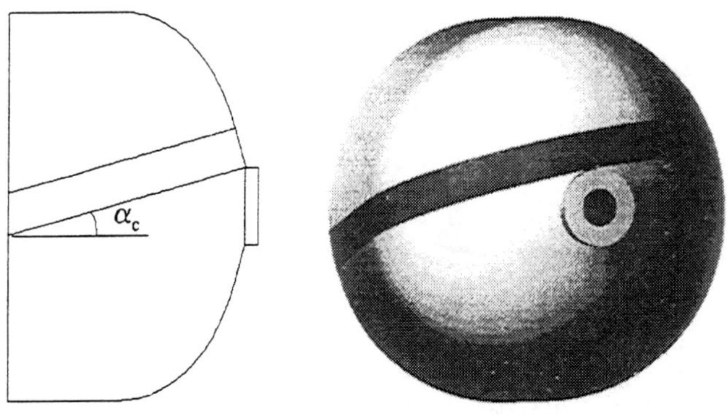

FIGURE 3 PLANAR WINDING PATTERN

have different opening sizes on the two ends. Unless the winding angle is to change along the length of the cylinder section, geodesic paths cannot be defined on both domes if the opening sizes are different.

A simple winding path to define and control is the *planar* winding pattern. For this pattern, the winding path forms a plane, as shown in Figure 3. For a certain type of winding machine, the polar winder, this is the only type of pattern that can be manufactured. Polar winders have become less popular in recent years, as they lack the versatility of computer-controlled, multi-axis helical winding machines. However, for high-quantity production of tanks, polar machines are inherently faster than helical winders, as the circular path of the winding head allows it to be run at a constant speed, while the head of a helical winder must decelerate, reverse direction, and accelerate in the opposite direction every time it reaches the end of a part.

The position of the fiber on the dome can be defined in polar coordinates by its longitudinal position x, radial position r, and angular position θ. The angle θ and its derivatives will be important in the determination of the orientation angle α and slippage tendency at any point. To find the angle θ in terms of r and x, consider the geometry of the winding pattern as shown in Figure 4. The depth of the dome d and the radius to the mid-line of the winding band r_p can be calculated from previously defined parameters and the width of the winding band BW and radius of the polar opening r_o:

$$r_p = r_o + \left(\frac{BW}{2}\right)\cos\phi_{cone} \qquad (10)$$

$$d = x_{cone} + \left(r_o + h_{cone} - r_p\right)\tan\phi_{cone} \qquad (11)$$

The coordinate y can be defined as:

$$y = r_p - (d - x)\tan\alpha_c \qquad (12)$$

or

$$y = r\sin\theta \qquad (13)$$

Equating Eqs. (12) and (13) and solving for θ:

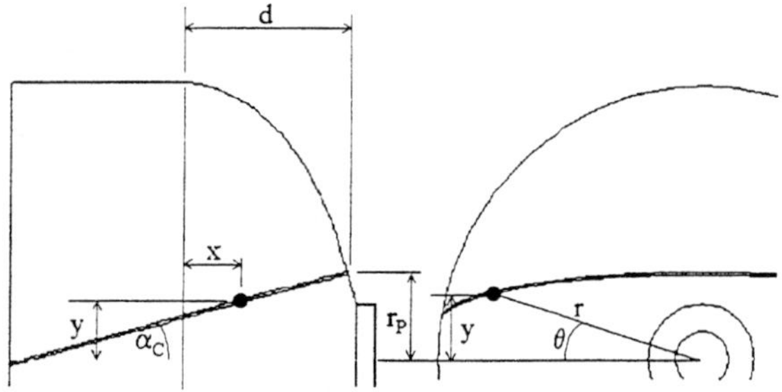

FIGURE 4 LOCATION OF FIBER PATH IN A PLANAR PATTERN

$$\theta = \sin^{-1}\left(\frac{r_p - (d - x)\tan\alpha_c}{r}\right) \tag{14}$$

Now the orientation angle α can be found. Consider and increment of the fiber path Δs, as shown in Figure 5. The angle at this location can be written as:

$$\tan\alpha = \frac{r\Delta\theta}{\sqrt{\Delta r^2 + \Delta x^2}} \tag{15}$$

Dividing numerator and denominator by Δx and taking the limit as Δx approaches zero:

$$\alpha = \tan^{-1}\left(\frac{r\theta'}{\sqrt{1 + r'^2}}\right) \tag{16}$$

The value of θ' can be found by differentiating Eq. (14):

$$\theta' = \frac{1}{\sqrt{1 - u^2}}u' \tag{17}$$

where

$$u = \frac{r_p - (d - x)\tan\alpha_c}{r} \tag{18}$$

and

$$u' = \frac{\tan\alpha_c - ur'}{r} \tag{19}$$

Now that the orientation angle of the fiber is known at every point on the dome contour, the local thickness can be calculated. The starting point will be the thickness in the cylinder region. In the design process, this will be the parameter that is varied until the stresses at any point in the dome contour are within acceptable limits. The cross-section area of the composite cylinder is:

$$A_{cyl} = \pi\left[\left(r_{cyl} + t_{cyl}\right)^2 - r_{cyl}^2\right] \tag{20}$$

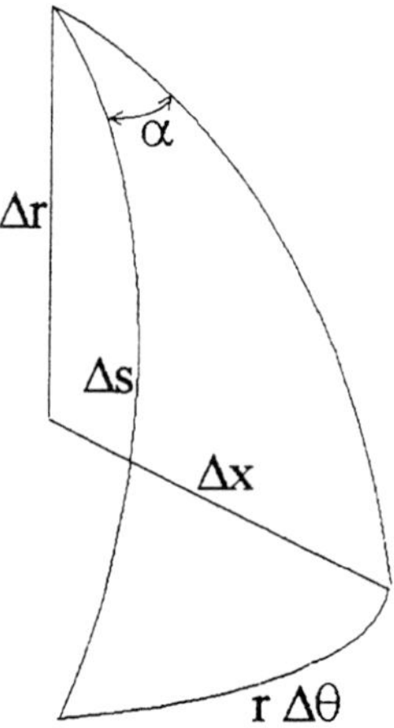

FIGURE 5 ORIENTATION ANGLE AT ANY POINT

Since the number of winding circuits (a circuit is a complete end-to-end and back path of the winding band) is constant with respect to axial location x, this is the parameter that will be used to calculate thickness at any point on the dome. In the cylinder, the area of a single winding band (cross section parallel to the tangent plane – see Figure 6) is

$$A_{band} = \frac{BW}{\cos\alpha_c}t_{band} \tag{21}$$

Therefore, the total cross-section of the composite in the cylinder is

$$A_{cyl} = \left(\#\text{ of bands}\right)A_{band} \tag{22}$$

Combining Eqs. (21) and (22) and solving for the number of bands,

$$\#\text{ of bands} = \frac{A_{cyl}\cos\alpha_c}{(BW)t_{band}} \tag{23}$$

For a point on the dome contour, the cross-section area of the composite is the surface of a conical segment. The area of this conical section is:

$$A = \pi t(2r + t\sin\phi) \tag{24}$$

As before, this area can be written in terms of the number of bands, the band width, and the local orientation angle:

$$A = \left(\#\text{ of bands}\right)\frac{BW}{\cos\alpha}t_{band} \tag{25}$$

Equating the two expressions for the area above, and substituting the expression for # of bands from Eq. (23), as the number of bands

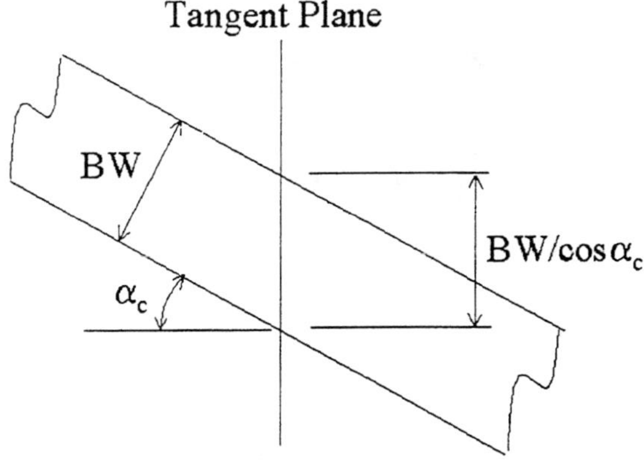

**FIGURE 6 BANDWIDTH PASSING THROUGH
TANGENT PLANE**

passing through any area must be the same, yield a quadratic equation for t. Solving for t:

$$t = \frac{-2\pi r + \sqrt{4\pi^2 r^2 + 4\pi \sin\phi\, A_{cyl}\left(\cos\alpha_c / \cos\alpha\right)}}{2\pi \sin\phi} \quad (26)$$

In summary, equations for the orientation angle and thickness at any point on the dome contour have now been developed.

Friction Calculations

Any deviation from a geodesic path requires that the friction between the liner and the composite (or between subsequent composite layers) be sufficient to prevent slippage. As noted earlier, this is a consideration even if the geodesic path is chosen, as the edges of the band will be following non-geodesic paths. Adding slippage calculations to the winding model allows for evaluation of feasible patterns, and also allows for optimization of the bandwidth. This last consideration is important for commercial applications, since manufacturing time depends on the amount of composite material that can be deposited in a given amount of time.

Alternatives to geodesic paths are the so-called semi-geodesic and quasi-geodesic paths (Liang et al, 1996). In the generation of these paths, a friction factor λ is assumed, and a path chosen so that the tangent component of the tension in the fibers is equal to the friction force λ times the normal force at any location. If $\lambda < |\mu|$, the coefficient of friction, then the path can be wound. The equation that is used to define the path is (Lossie and Van Brussel, 1994)

$$\frac{d\alpha}{ds} = \frac{\lambda\left[\left(1+r'^2\right)\sin^2\alpha - r\,r''\cos^2\alpha\right] - \left(1+r'^2\right)r'\sin\alpha}{r\left(1+r'^2\right)^{3/2}} \quad (27)$$

In the design approach formulated here, a path is assumed and the value of λ is calculated at every position for the edges of the winding band. Solving for λ,

$$\lambda = \frac{d\alpha/ds\left[r\left(1+r'^2\right)^{3/2}\right] + \left(1+r'^2\right)r'\sin\alpha}{\left(1+r'^2\right)\sin^2\alpha - r\,r''\cos^2\alpha} \quad (28)$$

Equations for r, r', and were developed. Therefore, the quantities dα/ds and r'' need to be evaluated to find the slip coefficient at every point on the dome contour. Since the other quantities have been found relative to the axial position x, the chain rule is utilized:

$$\frac{d\alpha}{ds} = \frac{d\alpha}{dx}\frac{dx}{ds} = \frac{\alpha'}{s'} \quad (29)$$

The value of α was found as Eq. (16). Differentiating with respect to x:

$$\alpha' = \frac{1}{1+v^2}v' \quad (30)$$

where

$$v = r\theta'\left(r'^2+1\right)^{-1/2} \quad (31)$$

and

$$v' = \frac{1}{\sqrt{r'^2+1}}\left[r\theta''+r'\theta' - \frac{r\,r'\,r''\theta'}{\left(r'^2+1\right)}\right] \quad (32)$$

and

$$r'' = -\frac{a}{b}\left[\left(b^2 - x^2\right)^{-1/2} + x^2\left(b^2 - x^2\right)^{-3/2}\right] \quad (33)$$

for the elliptical region (r'' = 0 for the conical region). To find θ'', Eq. (17) is differentiated:

$$\theta'' = \frac{1}{\sqrt{1-u^2}}\left(u'' + \frac{u\,u'^2}{\left(1-u^2\right)}\right) \quad (34)$$

where

$$u'' = \frac{1}{r}\left[\frac{u\,r'^2}{r} - \frac{\tan\alpha_c\,r'}{r} - u\,r'' - u'r'\right] \quad (35)$$

To find s', refer to Figure 5, where it can be seen that

$$\Delta s = \sqrt{r^2\left(\Delta\theta^2\right) + \left(\Delta x\right)^2 + \left(\Delta r\right)^2} \qquad (36)$$

Dividing both sides by Δx and taking limits as Δx approaches zero:

$$s' = \sqrt{1 + r^2\theta^2 + r'^2} \qquad (37)$$

Now Eq. (28) can be evaluated at any point along the dome profile, with λ the coefficient of friction required to prevent slippage of the band at that location.

Stress Resultants

Once the winding path has been established, an analysis of the membrane stresses in the dome can be performed. Jawad and Farr (1979) give equations for stress resultants in elliptical and conical shells under internal pressure loading. For the elliptical region,

$$N_\phi = \frac{pr_2}{2} \qquad (38)$$

and

$$N_\theta = pr_2 \frac{2r_1 - r_2}{2r_1} \qquad (39)$$

where

$$r_1 = \frac{a^2 b^2}{\left(a^2 \sin^2\phi + b^2\cos^2\phi\right)^{3/2}} \qquad (40)$$

and

$$r_2 = \frac{a^2}{\left(a^2 \sin^2\phi + b^2\cos^2\phi\right)^{3/2}} \qquad (41)$$

In the conical region, the stress resultants can be written as

$$N_\phi = \frac{pr}{2\sin\phi_{cone}} \qquad (42)$$

and

$$N_\theta = \frac{pr}{\sin\phi_{cone}} \qquad (43)$$

Classical lamination theory, as detailed by Jones (1975), is used to find the stresses in the fiber's local coordinate system.

DOME DESIGN METHODOLOGY

The equations developed in the previous sections form the basis of a design procedure for finding the best dome shape for a vessel. Figure 7 shows the steps in the calculation process. Input parameters include composite and liner material properties, liner thickness, cylinder diameter, polar opening diameter, and coefficient of friction. The coefficient of friction depends on the fiber, resin, and liner surface and is determined experimentally – Di Vita and Grimaldi (1990) give values from 0.1 to 0.3 for preimpregnated materials. Design variables include the ellipse ratio a/b, the winding angle in the cylinder, the

thickness of the cylinder material, the height of the cone, and the bandwidth. For this model, the cone height must be greater than the bandwidth.

An important feature of this model is that calculations are not performed only for the centerline of the winding band. Rather, slip calculations are performed for the edges of the band, and stress calculations are performed for ten subdivisions of the winding band. To illustrate the importance of this step, consider the variation in winding angle for the center of the band and the two outside divisions of the band illustrated in Figure 8. As the winding approaches the opening, the difference in winding angle across the band becomes large. This example is for a 1" wide band on a 20" diameter vessel, a reasonable pattern. When the strength calculations are performed (Figure 9), it is seen that considering the center of the band only results in a maximum stress calculation which is 12% less than that predicted by subdividing the band. Of course, this difference could be greater or less for other designs.Note that a large discontinuity in the stresses occurs at the ellipse-to-cone junction. Because the membrane stress resultants are not continuous in this region, the membrane stresses cannot be considered to be accurate. The addition of discontinuity analysis will allow these stresses to be analyzed.

A spreadsheet was created using this methodology. The spreadsheet allows for the selection of the best a/b ratio, winding angle, and thickness. Also, the bandwidth can be maximized for greater manufacturing speed. Finally, the cone height can be adjusted to yield a specific cone angle. This last feature allows for a common polar boss design to be used on different models of vessels.

DESIGN EXAMPLE

Consider a 50 cm (20 in.) diameter carbon-epoxy vessel with a 6 mm (0.25 in.) thick polyethylene liner. The design pressure is 10.3 Mpa (1,500 psi), and the polar opening in the dome is to be 10 cm (4 in.) in diameter. The design parameters shown in Table 1 are to be used.

TABLE 1 EXAMPLE DESIGN PARAMETERS

P	10.3 Mpa (1500 psi)
r_{cyl}	25 cm (10 in.)
r_o	5 cm (2 in.)
μ	0.15
Carbon-epoxy properties:	
E_1	124 GPa (18.0 X 10^6 psi)
E_2	6.9 GPa (1.0 X 10^6 psi)
ν_{12}	0.30
G_{12}	4.8 GPa (0.70 X 10^6 psi)
σ_1 Allowable	1.4 GPa (200,000 psi)
Polyethylene properties:	
E	2.1 GPa (300,000 psi)
ν	0.40
t_{liner}	6 mm (0.25 in.)

We will use a cone height of 5 cm (2 in.) for this example. This value can be adjusted later if desired. As mentioned earlier, adjustment of the cone height changes ϕ_{cone}, possibly allowing an existing polar boss to be used.

We begin by assuming an a/b ratio, in this case 1.5. A 25 mm (1 in.) wide winding band will be assumed. These parameters are entered into the design spreadsheet, and winding angle α_c is adjusted until the minimum slip coefficient λ is found. In this case, a winding

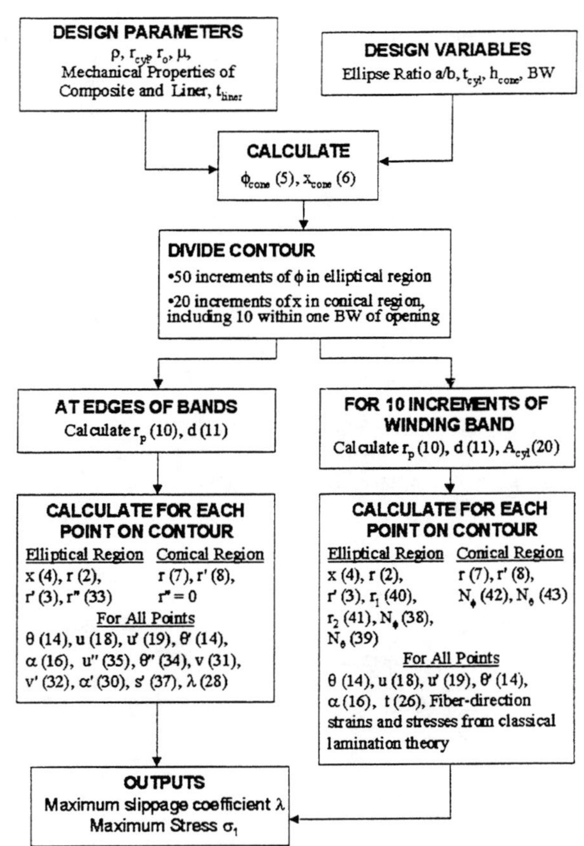

FIGURE 7 DOME DESIGN METHODOLOGY

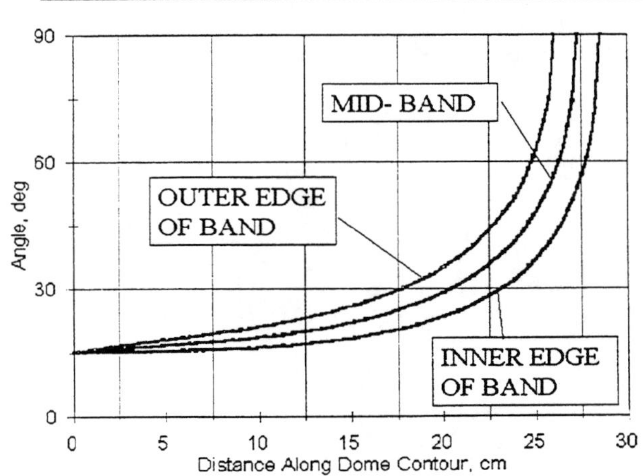

**FIGURE 8 WINDING ANGLE VARIATION THROUGH
WINDING BAND**

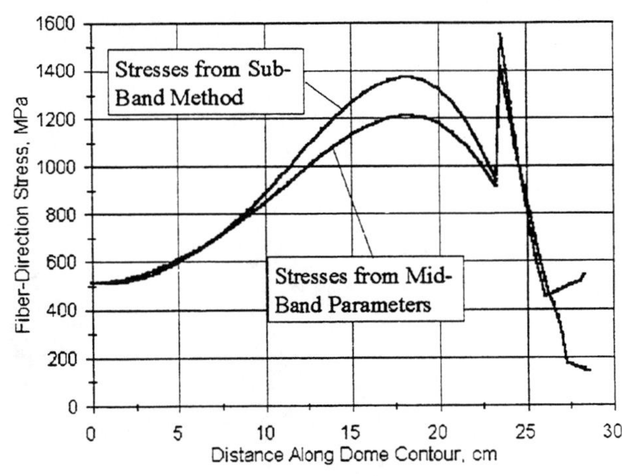

**FIGURE 9 COMPARISON OF STRESSES – SUB-
BAND METHOD vs. MID-BAND METHOD**

angle of 15° gives the lowest value of λ, 0.125. For comparison purposes, Figure 10 shows the slip coefficient at various locations for winding paths with angles of 10°, 15°, 20°, and 25°. The greatest tendency to slip occurs at the tangent plane, the ellipse-to-dome junction, or at the polar opening.

With the winding angle selected for minimum slip, the thickness is now chosen to meet the fiber stress criterion. In this case, a thickness of 2 mm (0.081 in.) results in a fiber-direction stress of slightly less than 1.4 GPa (200,000 psi). Another consideration in the design is the amount of hoop strain in the dome. Large negative strains are predicted near the tangent plane, raising the possibility of buckling in this region. As with metal vessels, the discontinuity stresses tend to lessen the severity of the compressive stresses in this region (Jawad and Farr, 1989). The discontinuity stress calculations will be added to the model in the future. Also, the hoop strains are relatively large further down the dome, where the fibers are still oriented more toward the meridian direction rather than the hoop direction. These large strains will result in non-catastrophic matrix cracking in the composite, but could also result in yielding or failure of the liner. A failure criterion for the liner also needs to be added to the model.

Table 2 shows the calculations made for different values of a/b ratio. As the ratio approaches 1.0 (spherical contour), the maximum hoop strain increases greatly. As the a/b ratio increases (flatter dome), the minimum hoop strain becomes greater in magnitude in compression, increasing the possibility of buckling.

TABLE 2 VARIATION OF SLIPPAGE TENDENCY AND STRAINS WITH a/b RATIO

a/b	Angle α_c	λ	t_c	Min. Hoop Strain	Max. Hoop Strain
1.0	17°	0.103	0.073 in.	0.2%	6.2%
1.25	15°	0.101	0.075 in.	0.2%	4.0%
1.50	15°	0.125	0.081 in.	-1.9%	3.2%
1.75	15°	0.146	0.090 in.	-7.2%	2.8%
2.00	15°	0.164	0.099 in.	-13.0%	2.7%

The "best" contour cannot be determined without considering issues such as envelope constraints (flatter domes can yield a greater tank

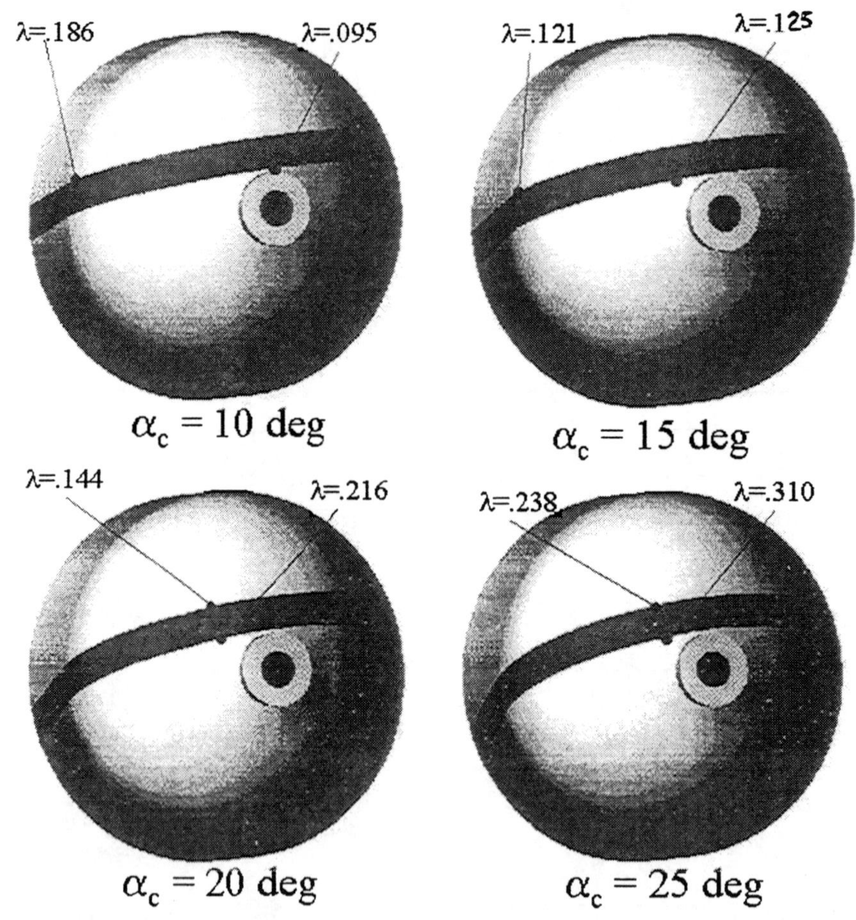

FIGURE 10 VARIATION OF SLIPPAGE TENDENCY WITH ORIENTATION ANGLE

volume in a length-restrained application), failure of the liner, or buckling of the dome.

Assuming that the a/b=1.5 dome is chosen, we can vary the bandwidth to predict the onset of slippage. In this case, the band can be increased to about a 38 mm (1.5 in.) width before the slip coefficient reaches the friction coefficient of 0.15. Also, the cone height can be adjusted to achieve a specific cone angle, if desired.

CONCLUSIONS

The design method presented here can be used to compare different dome contours, predict the tendency of the band to slip during manufacturing, and establish preliminary angles and thickness for a vessel design. The selection of elliptical–conical dome contours and planar winding patterns makes the method particularly suited for high-production-rate vessels.

REFERENCES

Di Vita, G. and Grimaldi, M., 1990, "The Filament Winding Manufacturing Technique: Studies on the Determination of the Friction Coefficient and on the Optimization of the Feed-Eye Motion," *22nd International SAMPE Technical Conference*, pp.972-977.

Fukunaga, H. and Uemura, M., 1983, "Optimum Design of Helically Wound Composite Pressure Vessels," *Composite Structures*, v. 1, pp.31-49.

Hojjati, M., Ardebili, V. Safavi, and Hoa, S. V., 1995, " Design of Domes for Polymeric Composite Pressure Vessels," *Composites Engineering*, v. 5, pp. 51-59.

Jawad, M., and Farr, J., 1989, *Structural Analysis and Design of Process Equipment, 2nd Edition*, John Wiley and Sons.

Jones, R. M., 1975, *Mechanics of Composite Materials*, McGraw-Hill.

Liang, Y. D., Zuo, Z. Q., and Zhang, Z. F., 1996, "Quasi-Geodesics - A New Class of Simple and Non-Slip Trajectories on Revolutional Surfaces," *28th Annual SAMPE Technical Conference Proceedings*, pp. 1071-1079.

Logan, D. L., and Hourani, M., 1983, "Membrane Theory for Layered Ellipsoidal Shells," *Transactions of the ASME- Journal of Pressure Vessel Technology*, v. 105, pp. 356-362.

Lossie, M. and Van Brussel, H., 1994, "Design Principles in Filament Winding," *Composites Manufacturing*, v. 5, pp.5-13.

Mistry, J. and Levy-Neto, F., 1994, "A Procedure for Analyzing the Strength of Composite Domes by Axisymmetric Shell Finite Elements," *Composites Engineering*, v.4, pp. 169-180.

ON CRACK GROWTH IN COMPOSITE CYLINDRICAL SHELLS

Hayder A. Rasheed
Bradley University
Peoria, Illinois 61625

John L. Tassoulas
The University of Texas at Austin
Austin, Texas 78712

ABSTRACT

Interfacial defects, in the form of cracks or layer separation, may occur in composite cylindrical shells during the manufacturing process, transportation or service life. Such defects are expected to affect the integrity of laminated composite structural elements and may reduce their capacity to resist the applied loads. In this article, the growth of pre-existing cracks in moderately thick composite cylinders is studied for the case of externally applied fluid pressure. The cracks considered separate thick layers, which are unlikely to buckle locally prior to the final collapse of the structural component. The potential of growth is assessed by computing the energy release rate. It is found that any initial out-of roundness imperfection introduces a shear force at the crack tip by causing the cross section to ovalize slightly. The energy release rate is found to vary exponentially with the applied pressure, when geometric nonlinearities are considered. The analysis is applied to a carbon/glass-fiber hybrid composite tube and the parameters influencing growth are examined. Crack length, through the thickness location, circumferential location relative to the ovalization orientation and the amount of imperfection are found to control the nature of growth. Unstable as well as stable crack growth and arrest cases are observed for various combinations of these parameters.

INTRODUCTION

The use of composites in constructing laminated cylindrical shells is increasing and expanding to a wide range of engineering applications. This includes aerospace, automotive, marine and civil engineering industries. In addition, the use of such structural components has been potentially considered for deep-water offshore drilling (Sparks1986). Composites suffer, like other laminated materials, from interfacial defects or layer separation which lead to loss of integrity. Such defects may pre-exist in composite cylindrical shells due to manufacturing imperfections (Kachanov 1974, Tarnopolskiy 1978), transportation impacts (Chai and Babcock 1985) or environmental effects (Sloan and Seymour 1990) during the service life of the structural component.

It has been observed that thin separated layers tend to locally buckle inwards, in rings and tubes under external pressure (Kachanov 1975). Few studies attempted to estimate the external pressure at the onset of local buckling of thin inner delaminated layers (Kachanov 1988, Chen and Simitses 1988). Such studies were limited to small prebuckling deformations, thin structual shells and isotropic materials or cross-ply symmetric composites. Recently, the effect of inner delamination buckling on the collapse pressure of orthotropic laminated composite tubes has been addressed in more detail (Rasheed and Tassoulas 1996a). Other studies considered the outer delamination buckling of thin separated layers in homogeneous isotropic or orthotropic cylinders (Marshall and Moshaiov 1990, Kardomateas and Chung 1992).

On the other hand, thick separated layers may have no tendency to buckle. However, such defects may still reduce the buckling capacity of structural shells, due to the considerable reduction in the flexural stiffness of the defective region.

Furthermore, the presence of a small initial out-of-roundness imperfection introduces a deviation from the uniform radial contraction under external pressure. This, in turn, introduces a shear force at the interfacial crack tip, which indicates a possibility of crack growth prior to the collapse of the cylindrical shell. This important issue has not been addressed in earlier studies. Accordingly, it is the subject of treatment in the present work.

This study investigates the potential of interfacial crack growth in long, fiber-reinforced laminated composite cylindrical shells subjected to external fluid-pressure loading. The crack is assumed to follow the lamination pattern and to extend along the cylinder. The finite element procedure developed takes into account the effects of crack-face contact, as well as initial imperfection, on the nonlinear geometric response. The energy release rate calculation is used to assess the contribution of the different defect parameters to crack growth.

GROWTH CRITERION

Cracks are likely to grow whenever the energy release rate assumes some critical value, (G_c). This seems to be the most widely accepted and used growth criterion for cracked composites (Whitcomb 1981, Chai and Babcock 1985, Kardomateas and Chung 1992, Lamborn and Schapery 1993).

$$G = G_c \qquad (1)$$

The critial value for the opening mode G_{Ic} is much smaller than that of the shearing mode G_{IIc} in conventional fiber-reinforced resin composites. In this case, it has been customary to substitute G_{Ic} for the mixed mode G_c in Eq.(1), to obtain conservative results (Chai and Babcock 1985, Sallam and Simitses 1985, Lamborn and Schapery 1993). Composites made out of toughened resins are reported to have high G_c values with comparable G_{Ic} and G_{IIc} levels (Bradley 1989). This allows Eq.(1) to yield more representative growth results, when G_{Ic} is used. In the present work, the total G, from analysis, is compared with G_{Ic} and G_{IIc} of non-toughened matrix composites, to establish lower and upper bound levels of pressure at growth. G is also compared with G_{IIc}, of toughened matrix composites. Typical G_c values are adopted for toughened and non-toughened matrix composites reinforced with glass and carbon fibers (Bradley 1989), Table 1.

FINITE ELEMENT MODEL

An orthotropic laminated shell finite element with a partially separated layer is developed for cross-section analysis of cylinders in generalized plane strain (Figure 1). Symmetry of the cross section, about a transverse axis, is assumed in the formulation, see Figure 1. The geometry and deformation of the outer shell is described by Fourier series expansions for radial and tangential displacements along the reference line, u_t and v_t, (Figure 2). Through the thickness in-plane rotation, ω_t, is also expanded by a similar Fourier series.

$$u_t = u_{t0} + a_{t1}\sin\theta + \sum_{n=2,4,\cdots}^{\infty} u_{tn}\cos n\theta + \sum_{n=3,5,\cdots}^{\infty} u_{tn}\sin n\theta \qquad (2)$$

$$v_t = -a_{t1}\cos\theta + \sum_{n=2,4,\cdots}^{\infty} v_{tn}\sin n\theta + \sum_{n=3,5,\cdots}^{\infty} v_{tn}\cos n\theta \qquad (3)$$

$$\omega_t = \sum_{n=1,3,\cdots}^{\infty} \omega_{tn}\cos n\theta + \sum_{n=2,4,\cdots}^{\infty} \omega_{tn}\sin n\theta \qquad (4)$$

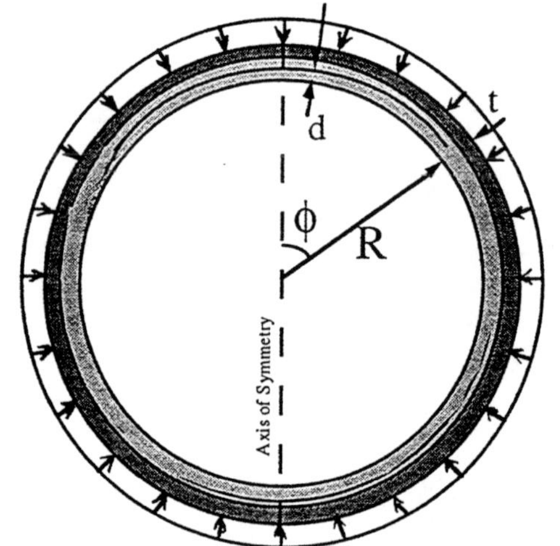

Fig.1 Composite cylinder with embedded crack.

The radial and tangential displacements are common between the inner and the outer shells (Figure 3). Through the thickness in-plane rotation of the inner shell, ω_t, is expressed by another Fourier series expansion, written in terms of the angle ϕ, see Figure 3 (Rasheed and Tassoulas 1996b).

$$\omega_{2t} = \omega_{2t\phi}\frac{\pi/2+\theta}{\pi-\phi} + \sum_{n=1}^{\infty}\omega_{2tn}\sin\frac{n\pi(\pi/2+\theta)}{\pi-\phi} \qquad (5)$$

The geometry of the reference line around the separated part is defined by Eqs.(2) and (3) to insure a smooth initial contact. The deformation of the separated part is also described by Fourier series expansions of radial and tangential displacements as well as through the thickness in-plane rotation as functions of the crack length (angle ϕ), (Figure 4).

$$u_s = u_{s\phi} + \sum_{n=1,3,\cdots}^{\infty} u_{sn} \sin \frac{n\pi\left(\phi + \hat{\theta}\right)}{2\phi} \qquad (6)$$

$$v_s = v_{s\phi}\frac{\hat{\theta}}{\phi} + \sum_{n=1}^{\infty} v_{sn} \sin \frac{n\pi\hat{\theta}}{\phi} \qquad (7)$$

$$\omega_s = \omega_{s\phi}\frac{\hat{\theta}}{\phi} + \sum_{n=1}^{\infty} \omega_{sn} \sin \frac{n\pi\hat{\theta}}{\phi} \qquad (8)$$

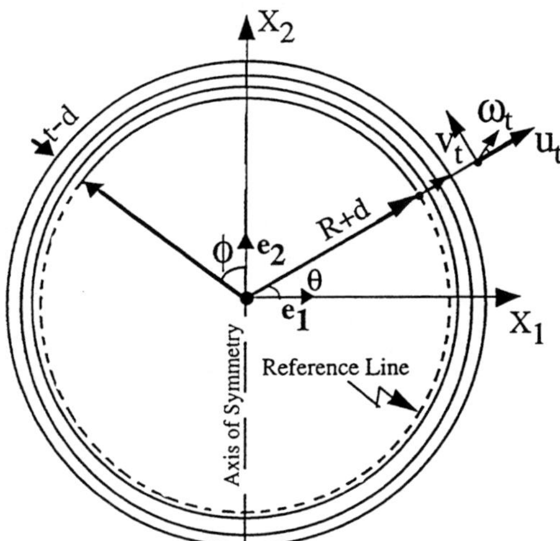

Fig.2 Outer shell part of the element.

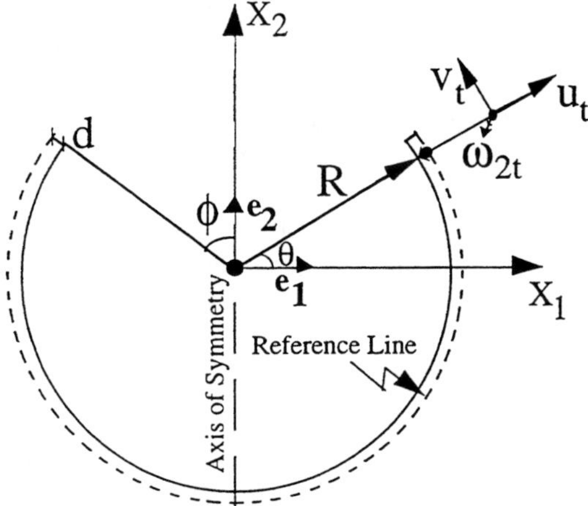

Fig.3 Inner shell part of the element.

The compatibility of deformation between the tube wall and the separated part, at the crack tip, is maintained by means of Lagrange multipliers. These correspond to three extra degrees of freedom introduced by imposing the following constraints:

$$u_t\left(\pi/2 - \phi\right) - u_s\left(\phi\right) = 0 \qquad (9)$$

$$v_t\left(\pi/2 - \phi\right) - v_s\left(\phi\right) = 0 \qquad (10)$$

$$\omega_{2t}\left(\pi/2 - \phi\right) - \omega_s\left(\phi\right) = 0 \qquad (11)$$

The degrees of freedom of the element are the out-of-plane strain ε and the coefficients of Fourier series in addition to the Lagrange multipliers.

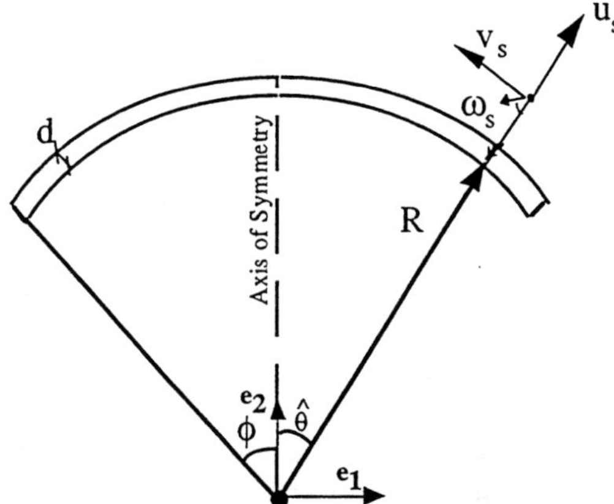

Fig.4 Separated part of the element.

Contact between the crack faces needs to be maintained as penetration takes place during the analysis. This generally occurs as a result of the inward deformation of the outer shell under external pressure. Frictional contact is assumed to be small and is neglected accordingly. Penetration normal to crack faces is prevented through a penalty formulation.

NUMERICAL PROCEDURE

Large tube deformation is considered in the formulation for the purpose of stability analysis under external pressure. The analysis assumes small initial out-of-roundness imperfections to the shell geometry. These imperfections are introduced by specifying initial values to the coefficients of the (2θ) terms of Fourier series to describe a slightly ovalized cylinder. This drives the solution into the first buckling mode shape.

In addition, the analysis algorithm searches for the bifurcation point of the local or delamination buckling by monitoring the eigenvalues of the element stiffness matrix. The detection of a bifurcation point is used to indicate the likelihood of such buckling. This is unlikely to occur for thick separated inner layers prior to the collapse of the cylindrical shell.

The nonlinear system of equilibrium equations is solved incrementally with Newton iterations performed at every step.

The tangent stiffness matrix and the internal load vector are derived using of the principle of virtual work.

$$\int_V \delta u_{i|j}\sigma^{ij}dV = \int_B \delta\mathbf{u}\cdot\mathbf{T}dB = \int_B \delta\mathbf{u}\cdot(-p\mathbf{n})dB \quad (12)$$

where σ^{ij} are the contravariant components of the Cauchy stress tensor, $\mathbf{T}$ is the boundary traction vector, p is the external pressure, $\mathbf{n}$ is the unit normal vector to the outer shell surface and $\delta\mathbf{u}$ is an arbitrary virtual displacement vector. The above equation is integrated over the reference configuration, V_0, and manipulated to obtain the incremental equilibrium equations (Rasheed and Tassoulas 1996b).

$$\int_{V_0} \delta U_{i|j}S^{ijkl}\Delta U_{k|l}dV_0 = \int_B \delta\mathbf{u}\cdot\mathbf{T}'dB -$$
$$\int_{V_0} \delta U_{i|j}(\mathbf{G}^i\cdot\mathbf{g}_k)\tau^{kj}dV_0 \quad (13)$$

where $\delta\mathbf{U}$ is the vector of arbitrary virtual degrees of freedom, S^{ijkl} are the components of the tensor of the tangent stiffness contribution including the material and geometric effects, $\mathbf{T}'$ is the boundary traction vector at a nearby equilibrium configuration, $\mathbf{G}^i$ are the contravariant base vectors in the reference configuration, $\mathbf{g}_k$ are the covariant base vectors in the current configuration, τ^{kj} are the contravariant components of the Kirchhoff stress tensor (Malvern 1969). The contact contributions to the left-hand and the right-hand sides of Eq.(13) are:

$$\int_{B_c} (\delta\mathbf{u_s} - \delta\mathbf{u_t})\cdot k_c\,(\Delta\mathbf{u_s} - \Delta\mathbf{u_t})\,dB_c \quad (14)$$

and

$$\int_{B_c} (\delta\mathbf{u_s} - \delta\mathbf{u_t})\cdot k_c\,(\mathbf{x_s} - \mathbf{x_t})\,dB_c \quad (15)$$

where B_c is the boundary of contact, k_c is the contact spring constant, $\mathbf{x_s} - \mathbf{x_t}$ is the penetration vector defined normal to the contact surface, $\delta\mathbf{u_s}$ and $\delta\mathbf{u_t}$ are the virtual displacement vectors for the separated layer and the tube wall at the contact surface. The constraints in Eqs.(9)-(11) are also added to Eq.(13) to form the complete incremental equilibrium equations (Rasheed and Tassoulas 1996b).

MATERIAL MODEL

A macromechanics-based composite material model was implemented. The model considers fiber-reinforced epoxy plies of different thickness, fiber orientation and fiber/matrix properties. Tsai-Wu and Hashin's ply failure criteria are adopted (Tsai and Wu 1971, Hashin 1980), with the option of using either one. Whenever failure in the transverse direction is indicated, transverse and in-plane shear moduli are reduced to zero at the local failure zone. Similarly, fiber and out-of-plane shear moduli are fully-degraded as failure in the ply fiber and out-of-plane shear are, respectively, reached.

ENERGY RELEASE RATE CALCULATION

The present formulation is based on the definition of the energy release rate as the derivative of the total potential energy with respect to the extension in the crack area. This derivative follows analogous steps to those of the formulation of the tangent stiffness matrix. Accordingly, G is derived for general nonlinear elastic cylinders. It is also applicable to inelastic cylinders when the proper secant moduli are used. This calculation is an extention to the stiffness derivative technique introduced by Parks (1974) for linear elastic systems:

$$G = -\frac{1}{2}\frac{\partial\pi}{\partial a} = -\frac{1}{2}\left(\frac{\partial U}{\partial a} - \frac{\partial W}{\partial a}\right) \quad (16)$$

where a is the crack length of half the cross section, due to symmetry, π is the total potential energy. The strain energy, U, and the potential of external loads, W, are defined as:

$$U = \frac{1}{2}\left[\int_{V_0} E_{ij}D^{ijkl}E_{kl}dV_0 + \right.$$
$$\left. \int_{B_{c0}} (\mathbf{x_s} - \mathbf{x_t})\cdot k_c(\mathbf{x_s} - \mathbf{x_t})dB_{c0}\right] \quad (17)$$

$$W = -\Delta v\,p \quad (18)$$

where E_{ij} are the covariant components of Green-Lagrange strain tensor, D^{ijkl} are the components of the elasticity tensor, Δv is the change in volume enclosed by the outer boundary under the pressure p. The model developed in this study allows the derivative of the potential energy to be explicitly evaluated with respect to the crack length in the perfectly-circular configuration. This can be easily transformed into its corresponding value with respect to the current configuration. Substituting Eqs.(17) and (18) into Eq.(16), simplifying the resulting derivative expressions using integration by parts and specializing those expressions to the case of generalized plane strain:

$$\frac{\partial U}{\partial a} = \int_{V_0} \delta U_{k|j}(\mathbf{G}^k\cdot\mathbf{g}_i)\tau^{ij}dV_0 +$$
$$\int_{B_{c0}} (\delta\mathbf{u}_s - \delta\mathbf{u}_t)\cdot k_c(\mathbf{x_s} - \mathbf{x_t})dB_{c0} +$$
$$\frac{1}{R+d}\left[\int_{V_0}(\mathbf{g}_i\cdot\frac{\partial\mathbf{g}_j}{\partial\phi} - \mathbf{G}_i\cdot\frac{\partial\mathbf{G}_j}{\partial\phi})\tau^{ij}dV_0 + \right.$$
$$\left. \int_{B_{c0}}\frac{\partial\mathbf{x}_s}{\partial\phi}\cdot k_c(\mathbf{x_s} - \mathbf{x_t})dB_{c0}\right] \quad (19)$$

$$\frac{\partial W}{\partial a} = \int_B \delta\mathbf{u}\cdot\mathbf{T}dB \quad (20)$$

It can be seen that the first two terms in Eq.(19), in addition to Eq.(20), satisfy the principle of virtual work. Accordingly, those terms vanish and an expression for G is reduced to:

$$G = -\frac{1}{2(R+d)} \left[\int_{V_0} (\mathbf{g}_i \cdot \frac{\partial \mathbf{g}_j}{\partial \phi} - \mathbf{G}_i \cdot \frac{\partial \mathbf{G}_j}{\partial \phi}) \tau^{ij} dV_0 + \int_{B_{c0}} \frac{\partial \mathbf{x}_s}{\partial \phi} \cdot k_c (\mathbf{x}_s - \mathbf{x}_t) dB_{c0} \right] \quad (21)$$

The above derivatives are written in terms of the crack length in the current configuration by simple mapping:

$$G = -\frac{1}{2} \frac{\partial \pi}{\partial a} \frac{da}{d\hat{a}} = -\frac{1}{2} \frac{\partial \pi}{\partial a} \frac{R+d}{\sqrt{g_{22}}} \quad (22)$$

where

$$\frac{da}{R+d} = \frac{d\hat{a}}{\sqrt{g_{22}}} \quad (23)$$

$\sqrt{g_{22}} = \sqrt{\mathbf{g}_2 \cdot \mathbf{g}_2}$, which is the length of the circumferential covariant base vector in the current configuration.

The above calculation was verified against an independant finite element program, adopting both the virtual crack closure and the stiffness derivative techniques and utilizing linear elastic four-node quadrilateral elements, allowing crack face overlapping (Rasheed and Tassoulas 1996b).

APPLICATIONS

IFP and Aerospatiale of France have designed, manufactured and tested long hybrid composite tube specimens made of high strength carbon/epoxy and S glass/epoxy layers (Sparks 1986). The specimens were not tested for crack defects. However, one of the specimen layups is adopted as a numerical prototype for studying crack growth, Figure 5. The tube has a 9 in. (229mm) inside diameter with a wall thickness of 0.606 in. (15.4mm) consisting of 13 alternately stacked glass and carbon-reinforced layers. The glass layers form 62% of the wall thickness, with their fibers running circumferentially at 90°. The carbon fibers are hellicaly wound at ±20° angle with the tube axis. Table 1 presents the mechanical properties of both layers.

THROUGH THE THICKNESS CRACK LOCATION

Thick separated layers are found to be unlikely to locally buckle prior to tube collapse. Accordingly, the main source of crack driving energy is the shearing force induced due to the ovalization imperfection and the variation in the flexural stiffness of the tube wall on both sides of the crack tip. Therefore, it is expected that G is mainly composed of G_{II}

knowing that the crack faces are modeled to be in smooth contact.

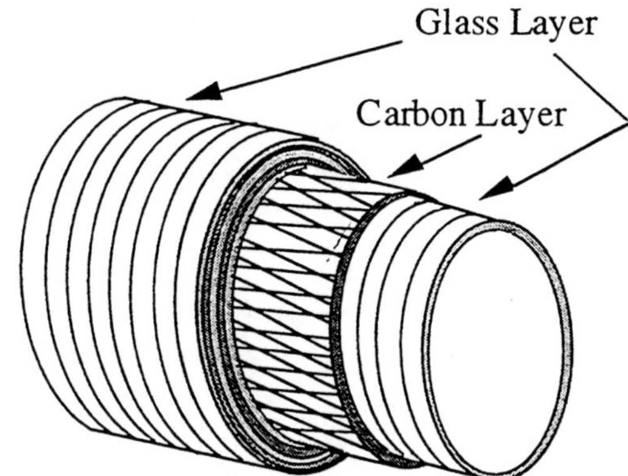

Fig.5 Composite tube specimen by IFP and Aerospatiale.

To demonstrate the above, a $\phi = 10°$ crack is assumed to pre-exist at the interface between the fourth (carbon) and fifth (glass) inner layers (d/t=0.3), see Figure 1. The cylinder is assumed to have an initial out-of-roundness imperfection of 0.4%. Figure 6 shows an exponential variation of G with external pressure. It also shows that G increases if the same crack is shifted to the the sixth-to-seventh layer interface (d/t=0.46), separating a thicker inner piece, Figure 6. The increase is attributed to the higher the shear stress expected at the crack tip in the second location. This supports the expectation that the fracture is mostly in mode II, for crack faces in perfect contact. However, gap imperfections may occur locally near the crack tip causing some mode I contribution. Comparisons with G_{Ic} and G_{IIc} indicate the likelihood of growth for this crack length prior to tube collapse, Figure 6. Comparison with G_{IIc} shows a potential of crack growth at a higher pressure level for the toughtned matrix composite, Figure 6.

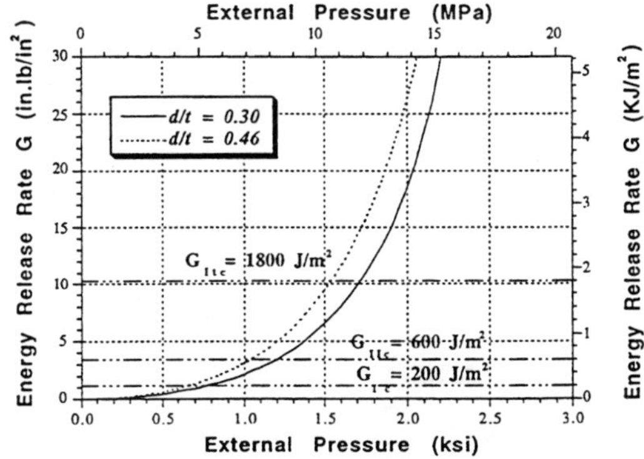

Fig.6 Energy release rate variation for $\phi=10°$.

CRACK LENGTH

Figures 7 demonstrates the same trend seen in Figure 6 for longer cracks (ϕ = 20-40°) situated at the second location (d/t=0.46). However, one may observe that the energy release rate is lower for longer cracks under the same pressure. This indicates a stable nature of crack growth in this range of crack length, since higher external pressure is needed to advance the longer cracks. Similar variation was found for the crack in the first location (d/t=0.3).

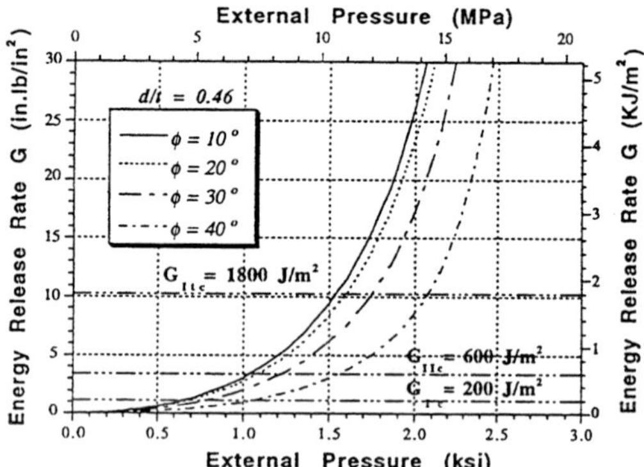

Fig.7 G variation for different crack lengths.

The above conclusions can be seen more clearly in Figure 8, plotted for d/t=0.46. It can be seen that the crack, in a toughened matrix, may extend from 20° to 32° by increasing the pressure from 1.60 ksi (11.04 MPa) to 1.80 ksi (12.42 MPa). It may also be observed that the crack is completely arrested at a crack angle of ϕ =47.5°. This is indicated by the zero G value obtained at this crack length, in the range of pressure presented in Figure 8. The numerically negative G values beyond this angle express no tendency to grow in the corresponding region, Figure 8.

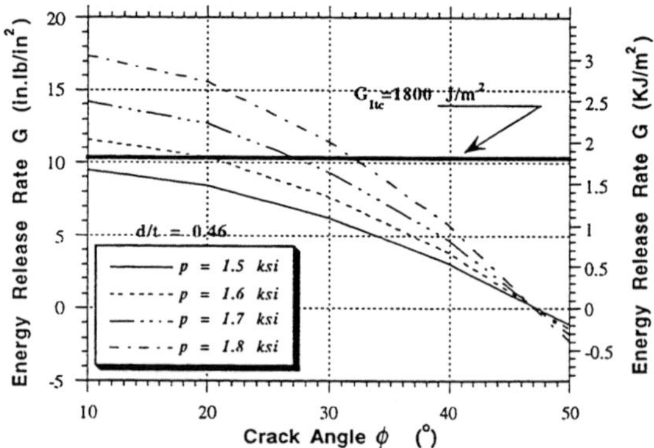

Fig.8 G variation with crack length.

To support this conclusion, the net cross section curvature and the shearing strain at the reference line are computed for points around the circumference at different levels of pressure. Figure 9 shows the net curvature variation along the circumference of the tube having ϕ =40°. This net curvature is assumed positive wherever the original tube curvature is increased. It can be seen that the inflection points are located at 49.5° and 125.5° for p=1.5ksi (10.35MPa). These locations are seen to shift to 44.5° and 126.5° for a crack of ϕ =50°, Figure 10.

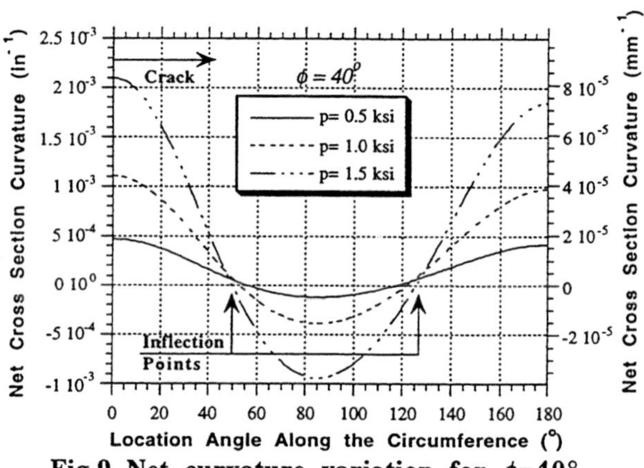

Fig.9 Net curvature variation for ϕ=40°.

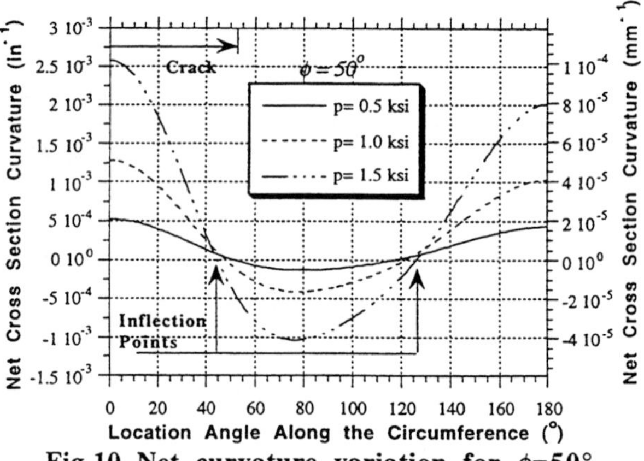

Fig.10 Net curvature variation for ϕ=50°.

Figures.11 and 12 present the shearing strain variation for the same cases above. It can be seen that the slope of the curves becomes zero at 42° and 135° angles for ϕ =40°. These positions adjust to 39° and 136° angles for ϕ =50°. G is seen to be positive and decreasing between ϕ =0° and 40°, where both parameters are positive and decreasing. It also becomes strictly negative wherever these parameters are both negative. This is attributed to the fact that G is the energy released at the crack tip during crack extension. This energy is a product of the shearing force and the relative sliding of the crack faces, for the case of mode II. Accordingly, G is a function of two terms, the first has the change of the shearing force with extension and the second has the change of the relative displacement with extension. The first term can be estimated from the slope of the shearing strain. The second term relates to the net curvature, which is the rate of change of the normal strain through the thickness.

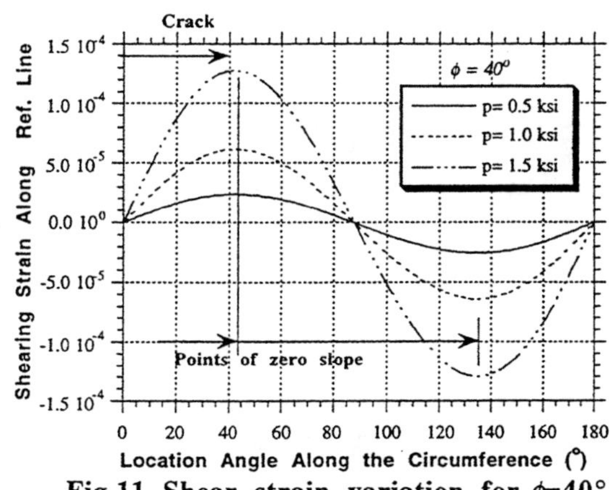

Fig.11 Shear strain variation for ϕ=40°.

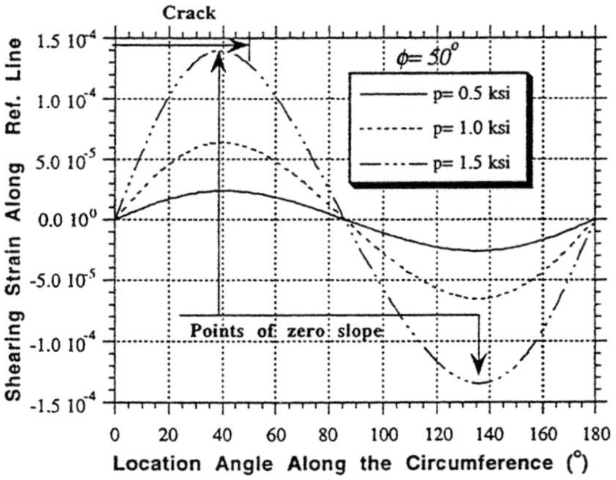

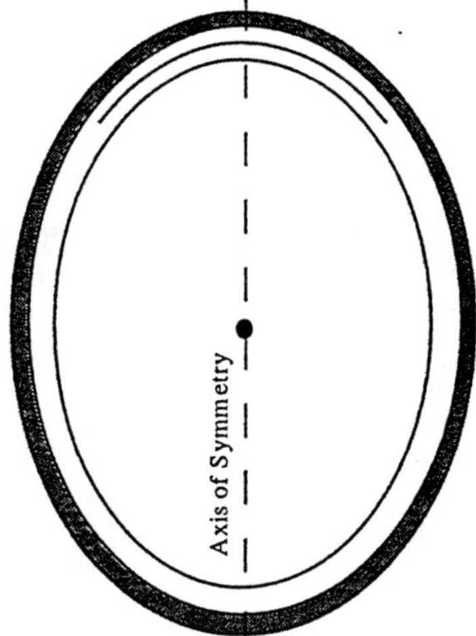

Fig.13 Ovalizing crack zone.

lengths. Figure 15 presents the variation of G with the crack angle at three pressure levels for the crack in the ovalizing region. It can be seen from this plot that the crack is possible to grow in the range of crack angles smaller than ϕ =48.5° and larger than ϕ =137°. It is also evident that the growth is stable for angles smaller than ϕ =48.5°and unstable for those larger than ϕ =137°. This is due to the fact that the crack tip is advancing away from the ovalizing zone, when $\phi < 48.5°$ and towards the ovalizing zone, for $\phi > 137°$.

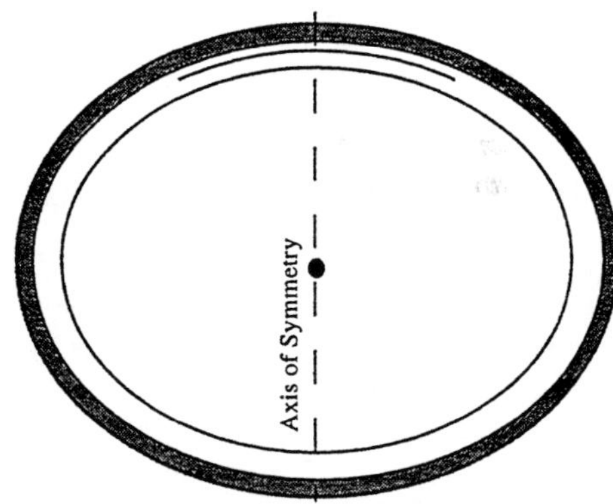

Fig.14 Flattening crack zone.

Fig.12 Shear strain variation for ϕ=50°.

CRACK LOCATION IN CROSS SECTION

The effect of crack location relative to the initial out-of-roundness imperfection is studied. The two extreme cases of having the crack centered at the zone of maximum and minimum initial curvature are considered, Figures. 13 and 14. These two locations are referred to as the ovalizing and flattening locations, respectively, since they tend to increasingly ovalize and flatten under external pressure. A crack of d/t=0.5 is assumed to pre-exist in each of the two locations and a parametric study is conducted to investigate the potential of growth for various crack

The same observations apply to the case of the flattening location with the crack possible to grow in the ϕ range of 46-135°, Figure 16. This behavior is attributed to the effects of the net curvature and the change in shearing strain along the circumference, as discussed in the previous section. These parameters cause the crack to have the tendency to grow in the ovalizing zone and close in the flattening zone.

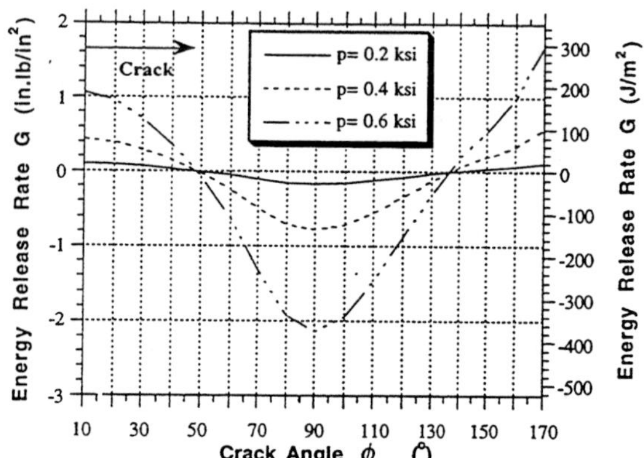

Fig.15 G variation for cracks in the ovalizing zone.

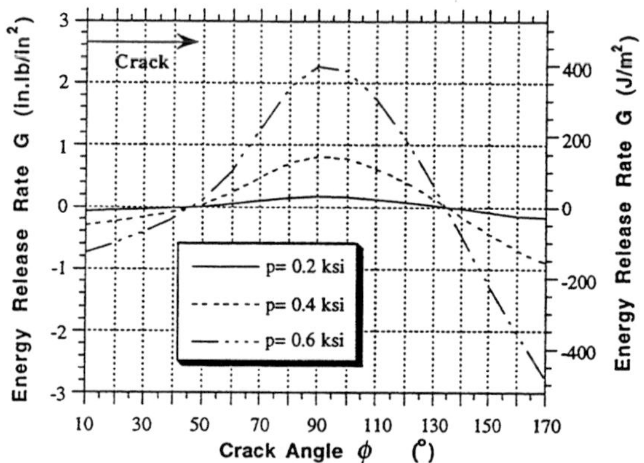

Fig.16 G variation for cracks in the flattening zone

INITIAL IMPERFECTION

The effect of initial imperfection on crack growth is also investigated. A crack with d/t=0.5 is selected to lie in the zone of maximum initial curvature, the ovalizing zone. Figure 17 presents the variation of G with external pressure for a crack angle of ϕ =20° and initial imperfections of 0.2, 0.4, 0.6 and 0.8%. It is evident that the potential of growth increases with the increase of this ovalization imperfection.

CONCLUSIONS

In this study, the stiffness-derivative technique is used to formulate the total energy release rate calculation for long laminated composite cylindrical shells having a pre-existing interfacial crack under external fluid-pressure. The formulation takes into account initial out-of-roundness imperfection, large deformations, crack face contact and material degradation. It is applied to study the potential of crack growth in a hybrid

composite tube. Thick separated layers are seen to have no tendency to locally buckle inwards until tube collapse. The energy release rate, for thick separated layers, is found to be mainly driven by the crack tip shear force induced due to the initial out-of-roundness imperfection. The growth is seen to be either stable, arrested or unstable depending on the length and location of the crack relative to the ovalization imperfection. The initial imperfection is shown to increase the likelihood of growth.

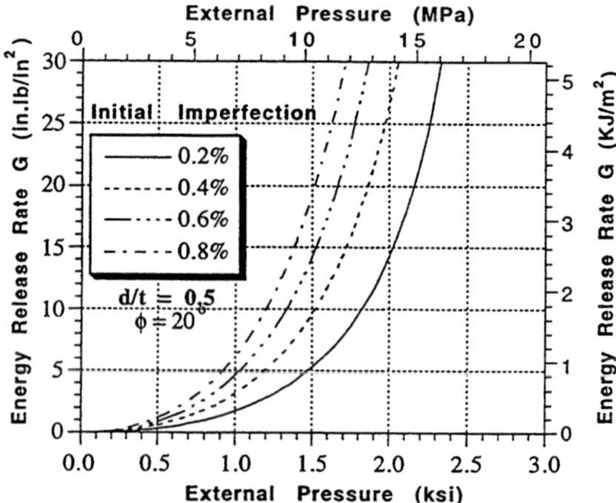

Fig.17 Effect of initial imperfection on growth.

ACKNOWLEDGEMENT

This work is part of a research project sponsored by the Offshore Technology Research Center, a National Science Foundation Engineering Research Center, under Grant CDR 8721512. The first author would also like to thank Prof. Richard A. Schapery for several valuable discussions during the course of this work.

REFERENCES

Bradley, W.L., 1989, "Relationship of Matrix Toughness to Interlaminar Fracture Toughness," *in* K. Friedrich (ed.): Application of Fracture Mechanics to Composite Materials (Vol.6), *part of* Composite Material Series, R.B. Pipes (series ed.),. *Elsevier Publishing Co.*, The Netherlands, 671pp.

Chen, Z.Q., and Simitses, G.J. 1988, "Delamination Buckling of Pressure-Loaded, Cross-Ply, Laminated Cylindrical Shells," *Zeitschrift feur Angewandte Mathematik und Mechanik (ZAMM),* Vol.68, No.10, pp.491-501.

Chai, H., and Babcock, C.D., 1985, "Two-Dimensional Modelling of Compressive Failure in Delaminated Laminates," *Journal of Composite Materials*, Vol.19, pp. 67-98.

Hashin, Z., 1980, "Failure Criteria for Unidirectional Fiber Composites," *Journal of Applied Mechanics*, ASME, Vol.47, pp.329-334.

Kachanov, L.M., 1974, "Cracks in Fiber-Glass Pipes," *Translated from Mekhanika Polimerov*, Vol.2, pp.370-371.

Kachanov, L.M., 1975, "Layering in Glass-Fiber Pipes Subject to External Pressure," *Translated from Mekhanika Polimerov*, Vol.6, pp.1106-1108.

Kachanov, L.M., 1988, "Delamination Buckling of Composite Materials," *Kluwer Academic Publishers*, Dordrecht, 95pp.

Kardomateas, G.A., and Chung, C.B., 1992, "Thin Film Modeling of Delamination Buckling in Pressure Loaded Laminated Cylidrical Shells," *AIAA Journal*, Vol.30, No.8, pp.2119-2123.

Lamborn, M.J., and Schapery, R.A., 1993, "An Investigation of the Existance of a Work Potential for Fiber-Reinforced Plastics," *Journal of Composite Materials*, Vol.27, No.4, pp.352-382.

Malvern, L.E., 1969, "Introduction to the Mechanics of a Continuous Medium," *Prentice-Hall*, Englewood Cliffs, NewJersy, 713pp.

Marshall, J.G. and Moshaiov, A., 1990, "The Buckling of Delaminated Rings Subjected to Hydrostatic Pressure," *Composite Structures*, Vol.14, pp.193-211.

Parks, D.M., 1974, "A Stiffness Derivative Finite Element Technique For Determination of Crack Tip Stress Intensity Factor," International Journal of Fracture, Vol.10, pp.487-502.

Rasheed, H.A., and Tassoulas, J.L., 1996a, "Effects of Delaminations on Composite Tubes Subjected to External Pressure," *Proceedings of the 6th International Offshore and Polar Engineering Conference*, ISOPE-96, Los Angeles, California, Vol.IV, pp.320-326.

Rasheed, H.A., and Tassoulas, J.L., 1996b, "Behavior and Strength of Composite Tubes Considering Delaminations and Other Defects," Offshore Technology Research Center, Austin, Texas, 191pp.

Sallam, S., and Simitses, G.J., 1985, "Delamination Buckling and Growth of Flat, Cross-Ply Laminates," *Composite Structures*, Vol.4, pp.361-381.

Sloan, F.E., and Seymour, R.J., 1990, "Environmental Testing Using A Compliant Load-Frame," *Journal of Composite Materials*, Vol.24, pp.727-738.

Sparks, C.P., 1986, ``Lightweight Composite Production Risers for a Deep Water Tension Leg Platform," *Proceedings, 5th International OMAE Conference*, Tokyo, Japan, pp.86-93.

Tarnopolskiy, Y.M., 1978, "A Delamination Failure Mode Of Composite Bars in Compression," *Proceedings, 1st US-USSR Symposium on Fracture of Composite Materials*, edited by Sih, G.C., and Tamuzs, V.P., pp.159-169.

Tsai, S.W., and Wu, E.M., 1971, "A General Theory of Strength of Anisotropic Materials," *Journal of Composite Materials*, Vol.5, pp.58-80.

Whitcomb, J.D., 1981, "Finite Element Analysis of Instability Related Delamination Growth," *Journal of Composite Materials*, Vol.15, pp.403-426.

Mechanical Property	Glass Layer	Carbon Layer
Fiber Content	66%	60%
$E_{11} in ksi(GPa)$	8300 (57.3)	20300 (140)
$E_{22} in ksi(GPa)$	2380 (16.4)	1160 (8)
$G_{12} in ksi(GPa)$	1117 (7.7)	580 (4)
$G_{23} in ksi(GPa)$	1117 (7.7)	580 (4)
$G_{31} in ksi(GPa)$	1117 (7.7)	580 (4)
ν_{12}	0.28	0.28
ν_{23}	0.42	0.42
ν_{13}	0.28	0.28
$S_{fiber}^{T} in ksi(MPa)$	232 (1600)	180 (1242)
$S_{fiber}^{C} in ksi(MPa)$	100 (690)	180 (1242)
$S_{matrix}^{T} in ksi(MPa)$	7.25 (50)	6 (41)
$S_{matrix}^{C} in ksi(MPa)$	20 (138)	25 (172)
$S_{shear} in ksi(MPa)$	12 (82)	12 (82)
$G_{Ic} in in.lb/in^2(J/m^2)$	1.43 (250)	1.14 (200)
$G_{IIc} in in.lb/in^2(J/m^2)$	11.40 (2000)	3.43 (600)
$G_{Itc} in in.lb/in^2(J/m^2)$	11.40 (2000)	10.28 (1800)

Table 1: Mechanical Properties of Composite Plies Used by IFP.

MD-Vol. 88, Polymeric Systems
ASME 1999

A STUDY ON THE CONSTITUTIVE LAW FOR POROUS SOLIDS WITH PRESSURE-SENSITIVE MATRICES AND ON THE MATERIAL BEHAVIOR OF RUBBER-TOUGHENED EPOXIES

Hyun-Yong Jeong
Department of Mechanical Engineering
Sogang University, Seoul, Korea
jeonghy@ccs.sogang.ac.kr

ABSTRACT

A macroscopic yield criterion for porous solids with pressure-sensitive matrices modeled by Coulomb's yield criterion was obtained by generalizing Gurson's yield criterion with consideration of the hydrostatic yield stresses for a spherical thick-walled shell and by fitting the finite element results of a voided cube. The macroscopic yield criterion is valid for negative mean normal stresses as well as for positive mean normal stresses. From the yield criterion, a plastic potential function for the porous solids was derived either for plastic normality flow or for plastic non-normality flow of pressure-sensitive matrices. In addition, elastic relations, an evolution rule for the plastic behavior of the matrices, a consistency equation and a void volume evolution equation were presented to complete a set of constitutive relations. The set of constitutive relations was implemented into a finite element code ABAQUS to analyze the material behavior of rubber-toughened epoxies. The cavitation and the deformation behavior were analyzed around a crack tip under three-point bending and around notch tips under four-point bending. In the numerical analyses, the cavitation of rubber particles was considered via a stress-controlled void nucleation model. The numerical results indicate that a reasonable cavitation zone can be obtained with void nucleation being controlled by the macroscopic mean normal stress, and a plastic zone is smaller around a notch tip under compression than under tension. These numerical results agree well with corresponding experimental results on the cavitation and plastic zones.

1. INTRODUCTION

In contrast to the classical plasticity theories, experiments showed that yielding of metals or polymers is dependent on the mean normal stress (Sternstein and Ongchin, 1969; Rabinowitz *et al.*, 1970; Sauer *et al.*, 1973; Spitzig *et al.*, 1975; Spitzig *et al.*, 1976; Spitzig and Richmond, 1979). The dependency of yielding on the mean normal stress, so called the pressure-sensitivity of yielding, has been modeled by Coulomb's yield criterion. However, if there exist microvoids or soft inclusions even in a solid which does not show any pressure-sensitivity of yielding, the macroscopic yielding of the solid becomes pressure-sensitive. Taking into account the macroscopic pressure-sensitivity of yielding, Gurson (1975, 1977) developed a yield function for porous solids.

It has been well known that some brittle polymers can be toughened by adding rubber particles. For example, rubber-toughened epoxies are ten times or more tougher than epoxy resins (Yee and Pearson, 1986; Pearson and Yee, 1986, 1991). Experiments showed that as rubber-toughened epoxies deform, rubber particles cavitate and shear yielding occurs subsequently (Yee and Pearson, 1986, Pearson and Yee, 1986, 1991). Thus, the cavitation can be considered as void nucleation occurring in the composite material of the pure epoxy matrix and rubber particles. In addition, in order to analyze the plastic deformation of the composite material both the pressure-sensitivity of the matrix and the macroscopic pressure-sensitivity of the composite material due to nucleated voids should be taken into account.

The author (1992; Jeong and Pan, 1995) and Lazzeri and Bucknall (1993) developed two different yield functions for porous solids with pressure-sensitive matrices, but these functions are not valid for negative mean normal stresses. However, in this paper a new yield function applicable to negative mean normal stresses as well as positive mean normal stresses was developed. From the yield criterion, a plastic potential function for the porous solids was derived either for plastic normality flow or for plastic non-normality flow of pressure-sensitive matrices. In addition, elastic relations, an evolution rule for the plastic behavior of the matrices, a consistency equation and a void volume evolution equation were presented to complete a set of constitutive relations. The set of constitutive relations was implemented into a finite element code ABAQUS to analyze the material behavior of rubber-toughened epoxies. The cavitation and the deformation behavior were analyzed around a crack tip under three-point bending and around notch tips under four-point bending, and they were compared with corresponding experimental results.

2. CONSTITUTIVE LAW

2.1. Yield Criterion

The pressure-sensitivity of yielding of steels or polymers has been modeled by Coulomb's yield criterion in which yielding is assumed to occur when a linear combination of the effective stress and the mean normal stress reaches the flow stress (Sternstein and Ongchin, 1969; Spitzig and Richmond, 1979; Kinloch and Young, 1983).

$$\tau_e + \mu \sigma_m = \tau_o \quad \text{or} \quad \sigma_e + \mu' \sigma_m = \sigma_o \qquad (1)$$

Here, τ_e is the effective shear stress, μ is the pressure-sensitivity factor, σ_m is the mean normal stress, τ_o is the shear flow stress, σ_e is the effective stress, μ' is $\sqrt{3}\mu$, and σ_o is $\sqrt{3}\tau_o$. Experiments showed that the pressure-sensitivity factor ranges from 0.014 to 0.064 for steels (Spitzig et al., 1975, 1976), and from 0.10 to 0.25 for polymers (Kinloch and Young, 1983). Moreover, the phase transformation of zirconia-containing ceramics can be also modeled by Coulomb's yield criterion, and the pressure-sensitivity factor for the phase transformation ranges from 0.55 to 0.93 (Yu and Shetty, 1989; Chen, 1991).

Gurson (1975, 1977) assumed a porous solid to have a periodic array of voids and to be an assembly of identical spheres with voids at the center. He also assumed the matrix to be pressure-insensitive, i.e. a von Mises material. Utilizing the upper bound approach, Gurson (1975, 1977) developed a yield function for the porous solid as follows.

$$\Phi_G(\Sigma,\sigma_o,f) = \left(\frac{\Sigma_e}{\sigma_o}\right)^2 + 2f\cosh\left(\frac{3\Sigma_m}{2\sigma_o}\right) - 1 - f^2 = 0 \qquad (2)$$

Here, Σ is the macroscopic Cauchy stress exerting on the porous solid, Σ_e is the macroscopic effective stress, Σ_m is the macroscopic mean normal stress, and f is the void volume fraction of the porous solid.

However, since it is difficult to find a kinematically admissible velocity field for the pressure-sensitive matrix modeled by Coulomb's yield criterion, the upper bound approach can not be used. Instead, the hydrostatic yield stress can be found, using the equilibrium equation and Coulomb's yield criterion, i.e. the lower bound approach, and it is expressed as follows.

$$(\Sigma_m)_y = \frac{\sigma_o}{\mu'}\{1 - f^{\left(\frac{\pm 2\mu'}{3 \pm 2\mu'}\right)}\} \qquad (3)$$

It can be proven from Eq. (3) that $(\Sigma_m)_y$ becomes equal to $\pm\frac{2}{3}\sigma_o \log f$ as μ approaches 0. That is, $2f\cosh\left(\frac{3\Sigma_m}{2\sigma_o}\right)$ becomes equal to $1 + f^2$, and Gurson's yield function is satisfied. In the same way, it can be also proven from Eq. (3) that $2f\cosh\left\{\frac{3+\text{sign}(\Sigma_m)2\mu'}{2\mu'}\log\left(1 - \mu'\frac{\Sigma_m}{\sigma_o}\right)\right\}$ becomes equal to $1 + f^2$ when μ is not equal to 0. In addition, a yield function for porous solids with pressure-sensitive matrices should reduce to Gurson's yield function when μ is equal to 0, and it should reduce to Coulomb's yield criterion when f is equal to 0. Based on these requirements, a yield function was proposed as follows.

$$\Phi_I(\Sigma,\sigma_o,f,\mu') = \left(\frac{\Sigma_e}{\sigma_o}\right)^2 + \left(1 - \mu'\frac{\Sigma_m}{\sigma_o}\right)^2 \left[2f\cosh\right.$$

$$\left.\left\{\frac{3+\text{sign}(\Sigma_m)2\mu'}{2\mu'}\log\left(1 - \mu'\frac{\Sigma_m}{\sigma_o}\right)\right\} - 1 - f^2\right] = 0 \qquad (4)$$

This yield function is better than the yield functions the author (1992; Jeong and Pan, 1995) and Lazzeri and Bucknall (1993) proposed because it is valid for negative mean normal stresses as well as positive mean normal stresses.

2.2. Interaction Between Voids

The yield function in Eq. (2) was obtained with no consideration of interaction between voids, and so was the yield function in Eq. (4). This results from the fact that Gurson (1975, 1977) assumed a porous solid to be an assembly of spherical shells, leaving space between them. Therefore, in order to take into account the interaction, it is necessary to adopt a voided cube instead of a spherical shell. However, since it is impossible to obtain the yield stresses of a voided cube analytically, in this paper the finite element method was used for a voided cube with the matrix being assumed to be almost rigid-perfectly plastic ($E/\sigma_o = 2\text{x}10^9$). Appling axisymmetric loading to the cube, the macroscopic mean normal stress and the macroscopic effective stress were obtained when massive plastic deformation occurred in the cube. The macroscopic stress invariants were plotted as empty symbols in Figs. 1 – 4 after being normalized by σ_o. As expected, the macroscopic stress invariants were located inside the loci defined by Eq. (4), which means that the interaction between voids helps the porous solid yield at lower loads. In order to fit better the FEM results, three parameters $(q_1 = 1.35, q_2 = 0.95, q_3 = 1.35)$ proposed by the author (1992) were also incorporated in Eq. (4) as follows.

$$\Phi_y(\Sigma, \sigma_o, f, \mu') = \left(\frac{\Sigma_e}{\sigma_o}\right)^2 + \left(1 - \mu'\frac{\Sigma_m}{\sigma_o}\right)^2 \left[2q_1 f \cosh\right.$$

$$\left\{q_2 \frac{3 + \text{sign}(\Sigma_m)2\mu'}{2\mu'} \log\left(1 - \mu'\frac{\Sigma_m}{\sigma_o}\right)\right\} - 1 - q_3 f^2 \bigg] = 0 \quad (5)$$

However, Tvergaard (1981, 1982) introduced different values of parameters $(q_1 = 1.5, q_2 = 1.0, q_3 = 2.25)$ into Gurson's yield function by comparing the plastic flow localization results of his finite element computations with those of a continuum model based on Gurson's yield function.

2.3. FEM Results and Modified Yield Function

In Figs. 1 - 4, the yield function Φ_y (curves) along with the FEM results (empty symbols) and the hydrostatic yield stresses $(\Sigma_m)_y$ (solid symbols) are shown for four different void volume fractions and for four different pressure-sensitivity factors. It is noteworthy that the yield function correlates well with the FEM results for a wide range of the void volume fraction and the pressure-sensitivity factor. The difference between the solid symbols and the empty symbols located on the abscissa shows the effect of interaction between voids on yielding. Figures 1 - 4 show that the difference increases as f or μ increases, especially for the negative hydrostatic stresses.

2.4. Plastic Potential Function

Experiments showed that during plastic deformation steels or polymers modeled by Coulomb's yield criterion dilate less than is predicted by the normality plastic flow rule (Spitzig et al., 1975, 1976; Spitzig and Richmond, 1979). This means that the plastic potential function is not the same as the yield function. If the dilatancy factor is denoted by β and the effective shear plastic strain is denoted by γ_e^p, the volume increase during plastic deformation can be expressed as $\beta\gamma_e^p$, and the plastic potential function can be defined as follows.

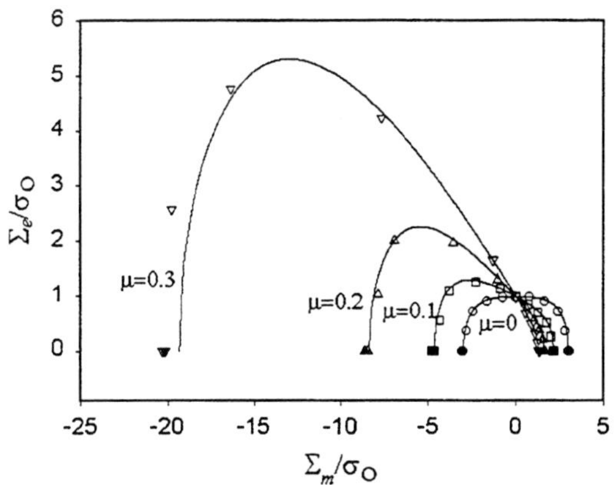

Fig. 1. $(\Sigma_m)_y$, FEM results, and the yield function for $f = 0.01$

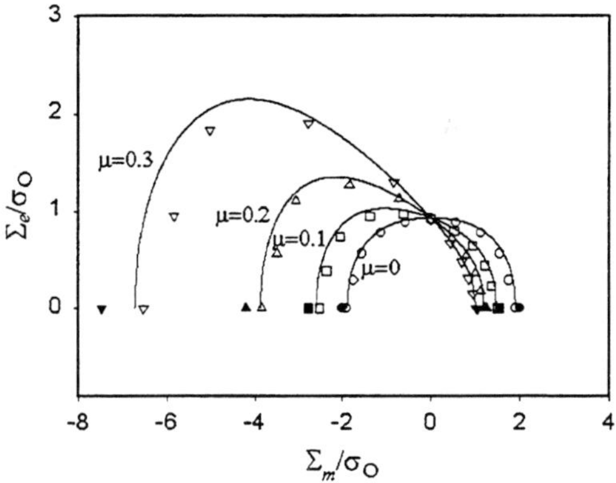

Fig. 2. $(\Sigma_m)_y$, FEM results, and the yield function for $f = 0.05$

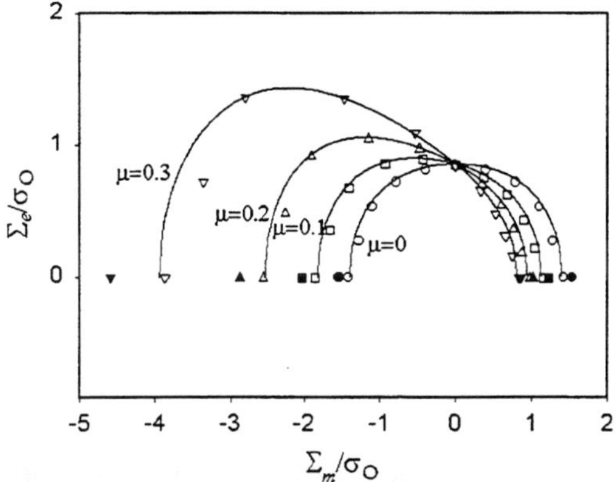

Fig. 3. $(\Sigma_m)_y$, FEM results, and the yield function for $f = 0.10$

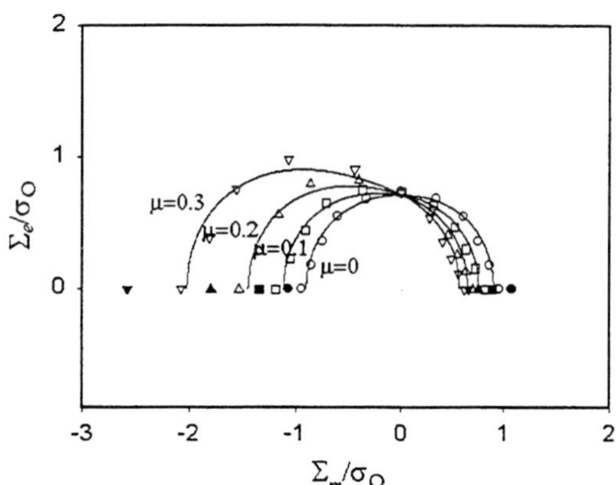

Fig. 4. $(\Sigma_m)_y$, FEM results, and the yield function for $f = 0.20$

$$\tau_e + \beta\sigma_m = \tau_p \qquad (6)$$

Note that the plastic potential function is the same as the yield function except that μ and τ_o are replaced with β and τ_p, respectively. Here, τ_p is the fictitious shear flow stress that makes the current stress state satisfy Eq. (6). Therefore, the plastic potential function for porous solids with pressure-sensitive matrices can be obtained in the same way. That is, replacing μ' with $\beta'(=\sqrt{3}\beta)$ and σ_o with σ_p in the yield function $\Phi_y(\Sigma,\sigma_o,f,\mu')$ results in the plastic potential function $\Phi_p(\Sigma,\sigma_p,f,\beta')$.

2.5. Elastic Relations

When the void volume fraction is large, its effect on elastic moduli should be taken into account. Adopting the self-consistent scheme combined with the average stress scheme (Tandon and Weng, 1988), elastic moduli were calculated from those for matrix and the void volume fraction as follows.

$$E^* = \frac{2E(7-5v)(1-f)}{14-10v+f(1+v)(13-15v)} \qquad (7)$$

$$v^* = \frac{v(14-10v)+f(1+v)(3-5v)}{14-10v+f(1+v)(13-15v)} \qquad (8)$$

In modeling rubber-toughened epoxies, E and v are elastic moduli of the composites before cavitation, and f is the volume fraction of cavitated voids.

2.6. Void Nucleation Model and Void Volume Evolution Equation

The increase of void volume arises both from the growth of existing voids and from the nucleation of voids. For the nucleation of voids, two models have been used; one is the plastic strain-controlled nucleation model and the other is the stress-controlled nucleation model. In this paper, only the stress-controlled nucleation model was used. For the stress-controlled nucleation model the sum of the tensile flow stress and the macroscopic mean normal stress was used as the controlling stress in many researches (for example, Tvergaard, 1982b; Pan *et al.*, 1983; Needleman and Tvergaard, 1987). The sum was suggested to be an approximation to the maximum stress transmitted across the particle-matrix interface, and void nucleation was believed to depend on the sum (Argon and Im, 1975). In this paper this stress-controlled nucleation model is named the first stress-controlled nucleation model. Since experimental results showed that the cavitation zones were circular and similar to the contour plots of the macroscopic mean normal stress (Pearson and Yee, 1991; Yee *et al.*, 1993), only the macroscopic mean normal stress can be also used as the controlling stress. This stress-controlled nucleation model is named the second stress-controlled nucleation model. Therefore, the void volume evolution equation can be expressed as follows.

$$\dot{f} = (1-f)[tr(D^P) - \beta'\dot{\varepsilon}_e^P] + A\dot{\sigma}_o + B\dot{\Sigma}_m \qquad (9)$$

Here, $\dot{\varepsilon}_e^P$ is the effective plastic strain rate, and $tr(D^P)$ is the trace of rate-of-deformation tensor. Moreover, for the first stress-controlled nucleation model,

$$A = B = \frac{f_B}{s\sigma_y\sqrt{2\pi}}\exp\left[-\frac{1}{2}\left(\frac{\sigma_o+\Sigma_m-\sigma_N}{s\sigma_y}\right)^2\right] \qquad (10)$$

However, for the second stress-controlled nucleation model

$$A = 0, \; B = \frac{f_B}{s\sigma_y\sqrt{2\pi}}\exp\left[-\frac{1}{2}\left(\frac{\Sigma_m-\sigma_N}{s\sigma_y}\right)^2\right] \qquad (11)$$

Here, f_B is the volume fraction of void nucleating particles, s and σ_N are the standard deviation and the mean value of the normal distribution function, respectively. In addition, voids nucleate only when the controlling stress is bigger than its maximum value occurred in the previous deformation history.

2.7. Strain Rate Sensitivity and Strain Softening-and-Hardening

Some polymers show initial intrinsic strain softening and subsequent hardening as well as strain-rate sensitivity. A simple power law with the strain-rate hardening exponent m was adopted as follows.

$$\dot{\varepsilon}_e^p = \dot{\varepsilon}_r \left[\frac{\sigma_o}{g(\varepsilon_e^p)} \right]^{1/m} \qquad (12)$$

where $\dot{\varepsilon}_r$ is the reference tensile plastic strain rate, and the function $g(\varepsilon_e^p)$ equals the tensile flow stress σ_o of the matrix when $\dot{\varepsilon}_e^p = \dot{\varepsilon}_r$. To model qualitatively the initial strain softening and subsequent hardening behavior of epoxies, a function for $g(\varepsilon_e^p)$ was proposed as follows.

$$g(\varepsilon_e^p) = \sigma_y \left[\left(\frac{\varepsilon_e^p}{\varepsilon_y} + 1 \right)^N + C_1 \left(\frac{\varepsilon_e^p}{\varepsilon_y} \right)^{N_1} \log \left(C_2 \frac{\varepsilon_e^p}{\varepsilon_y} \right) \right] \qquad (13)$$

where N is the hardening exponent, N_1 is a softening-hardening exponent, and C_1 and C_2 are two material coefficients. Here, σ_y is equal to $\sqrt{3}\tau_y$, where τ_y is the shear yield stress in shear at the reference shear plastic strain rate $\dot{\gamma}_r$ $(= \sqrt{3}\dot{\varepsilon}_r)$, where ε_y is σ_y/E. When C_1 is equal to zero, Eq. (13) reduces to the usual power-law strain hardening relation.

For detailed derivations of the constitutive law except the yield function and explanations on the meaning of the material properties, refer to other papers (Jeong, 1992; Jeong and Pan, 1995, 1996).

3. MATERIAL BEHAVIOR OF RUBBER TOUGHEND EPOXIES

3.1. Crack-Tip Model and Notch-Tip Model

The constitutive law developed in the previous sections was implemented as a user material subroutine in ABAQUS. As shown in Figs. 5 and 6, inner and outer meshes were created only for the upper half around a crack under Mode I plane strain loading. The outer radius of the outer mesh is 10,000 times bigger than the crack-tip radius, and the small-scale yielding condition was guaranteed. CPE8R (8 node plane strain reduced integration scheme) elements were used. This model was developed to simulate the material behavior of a rubber-toughened epoxy, DER331/Pip/CTBN-8(10), around a crack tip, which was experimentally investigated by Pearson and Yee (1986, 1991).

In addition, as shown in Fig. 7, a symmetric double-edge double-notched (SDEDN) specimen under four-point bending was meshed only for the left half, using 3-node and 4-node plane strain elements. The arrow in Fig. 7 shows the loading direction along which the bending crosshead moves. This model was developed to simulate the material behavior of the rubber-toughened epoxy around notch tips, which was experimentally investigated by Yee *et al.* (1993). Note that the material around the upper notch was under

compression, and the material around the lower notch was under tension.

The material properties for the rubber-toughened epoxy used in this paper are $E = 2530$ MPa, $v = 0.428$, $\sigma_y = 75$ MPa, $m = 0.035$, $N = 0.1$, $N_1 = 1.3$, $C_1 = 0.03$, $C_2 = 0.05$, $\mu = 0.13$, $\beta = 0$, and $f_N = 0.12$. For the first stress-controlled nucleation model, $\sigma_N = 1.0$ and $s = 0.25$ were used. However, for the second stress-controlled nucleation model, $\sigma_N = 0.75$ and $s = 0.25$ were used.

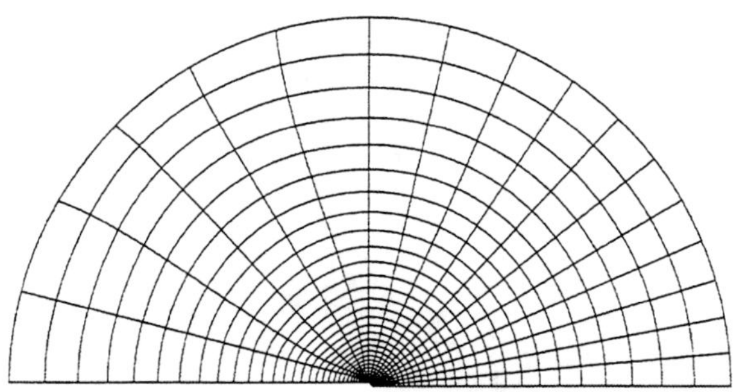

Fig. 5. Inner mesh around the crack tip

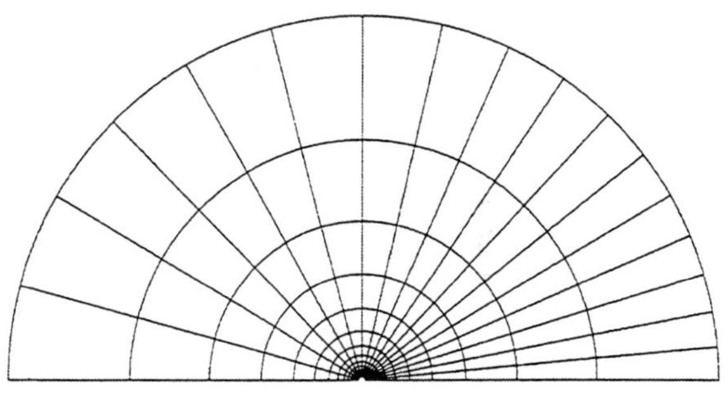

Fig. 6. Outer mesh around the crack tip

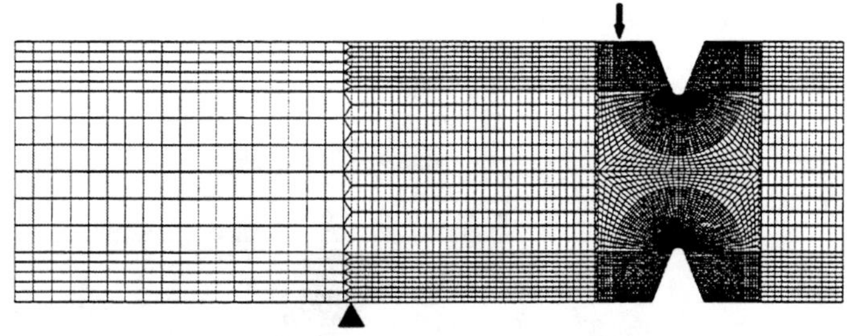

Fig. 7. FEM mesh for the SDEDN specimen

81

3.2. Material Behavior around the Crack Tip

From the numerical simulations the contour plots of void volume fraction were obtained when the crack opening displacement (COD) was about three times the original COD, and they are shown in Figs. 8 and 9, the coordinates being normalized by $(K_I / \sigma_y)^2$.

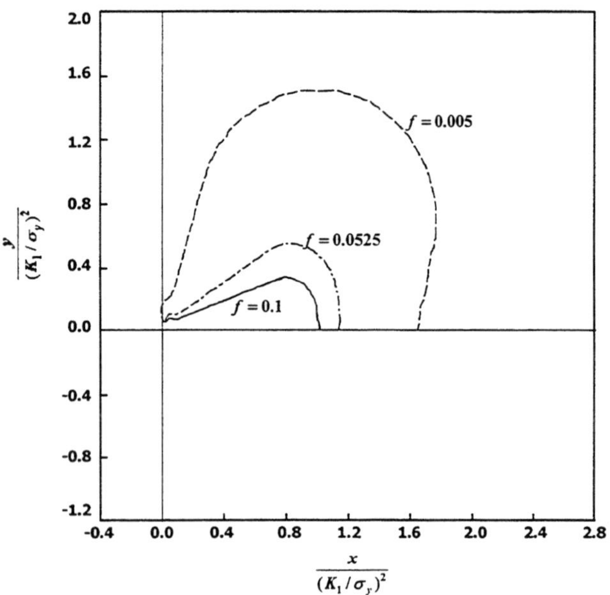

Fig. 8. Contour plots around the crack tip
(first stress controlled nucleation model)

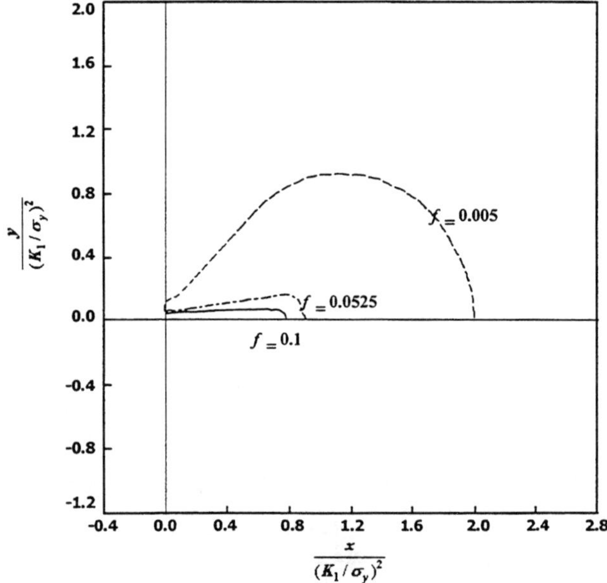

Fig. 9. Contour plots around the crack tip
(second stress controlled nucleation model)

The first stress-controlled nucleation model was used for Fig. 8, but the second stress-controlled nucleation model was used for Fig. 9. In both cases, the contour plot of $f = 0.005$ surrounds the crack tip, but the contour plot of $f = 0.1$ is located along the crack line. It is noteworthy that the contour plot of $f = 0.005$ shown in Fig. 8 is elongated along the major principal stress direction, but the contour plot of $f = 0.005$ shown in Fig. 9 is quite circular.

An optical micrograph near the crack tip in a rubber-toughened epoxy, DER 331/Pip/CTBN-8(10), was provided by Pearson and Yee (1991). It showed that a circular cavitation zone surrounded a shear yielding zone, and its diameter was about $1600 \, \mu m$. It also showed that the shear yielding zone was baseball-bat shaped, and it was formed along the crack line with the length of about $600 \, \mu m$. Since the fracture toughness, K_{IC}, for the rubber-toughened epoxy was reported to be 2.10 Mpa $\sqrt{m}$ (Pearson and Yee, 1991), the normalized diameter of the cavitation zone and the normalized length of the shear yielding zone are 2.04 and 0.77, respectively. Note that the contour plots of $f = 0.005$ and $f = 0.1$ shown in Fig. 9 have almost the same shape and size as the cavitation and the shear yielding zones observed in the optical micrograph.

3.3. Material Behavior around the Notch Tips

The contour plots of void volume fraction and effective plastic strain were obtained from the numerical simulations when the crosshead moved downward by $1 \, mm$, and they are shown in Fig. 10.

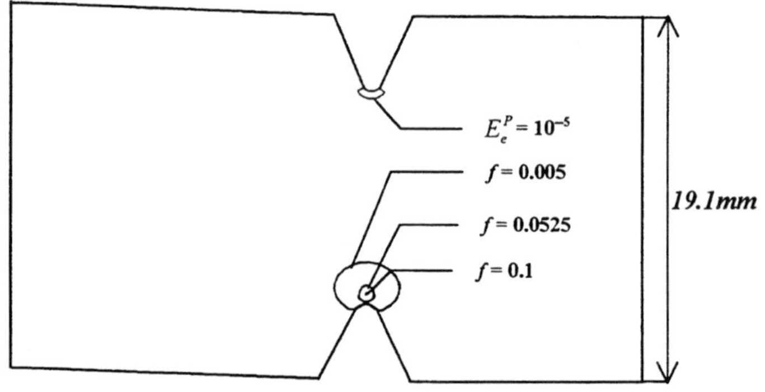

Fig. 10. Contour plots in the SDEDN specimen

The contour plots of void volume fraction are shown only around the lower notch (because no cavitation occurred around the upper notch), and the contour plot of effective plastic strain are shown only around the upper notch. Note that the plastic zone around the upper notch under compression is quite small, and the contour plot of $f = 0.005$ around the lower notch is quite circular. The diameter of the contour plot of $f = 0.005$

is about $2.6\,mm$. Experiments also showed that a circular cavitation zone occurred only around the lower notch, and the plastic zone was smaller around the upper notch under compression (Yee *et al.*, 1993). However, the experiments showed that the diameter of the cavitation zone was about $2\,mm$ when the crosshead moved downward about $1.5\,mm$. This discrepancy might have come from the uncertainty of the standard deviation and the mean value used for the void nucleation model. For Fig. 10, the second stress-controlled nucleation model was used. However, when the first stress-controlled nucleation model was used, the contour plot of $f = 0.005$ was elongated along the major principal stress direction as shown in Fig. 8, which is different from the shape of the cavitation zone observed in experiments.

4. CONCLUSIONS

The yield function developed in this paper is valid for negative mean normal stresses as well as the positive mean normal stresses, and it resulted in a reasonable plastic zone around the notch under compression. In addition, if the contour plot of $f = 0.005$ is regarded as a cavitation zone, it can be mentioned that the second stress-controlled nucleation model resulted in circular cavitation zones around a crack tip and around a notch tip. The cavitation zones obtained from numerical simulations, especially the cavitation zone around a crack tip, are quite similar to those observed in experiments in terms of shape and size. This means that the macroscopic mean normal stress may be the controlling stress for cavitaion in the rubber particles in rubber-toughened epoxies. However, a more research needs to be done on how to better model the cavitation process. If a normal distribution function is to be used for cavitation (or void nucleation) modeling, it is necessary to set up a methodology on how to better determine the values of the mean value and the standard deviation of the normal distribution function.

Acknowledgements

The financial support from Sogang University for this research is highly appreciated.

REFERENCES

Argon, A. S. and Im, J., "Separation of Second Phase Particles in Spheroidized 1045 Steel, Cu-0.6pct Cr Alloy, and Maraging Steel in Plastic Straining," *Metallurgical Transactions*, Vol. 6A, pp. 839-851.

Chen, I.-W., 1991, "Model of Transformation Toughening in Brittle Materials," *Journal of American Ceramic Society*, Vol. 74, pp. 2564-2572.

Gurson, A. L., 1975, "Plastic Flow and Fracture Behavior of Ductile Materials Incorporating Void Nucleation, Growth and Interaction," Ph.D. Thesis, Brown University.

Gurson, A. L., 1977, "Continuum Theory of Ductile Rupture by Void Growth: Part I-Yield Criteria and Flow Rules for Porous Ductile Media," *Journal of Engineering Materials Technology*, Vol. 99, pp. 2-15.

Kinloch, A. J. and Young, R. J., 1983, *Fracture Behavior of Polymers*, Elsevier Applied Science.

Jeong, H.-Y., 1992, "A Macroscopic Constitutive Law for Porous Solids with Pressure-Sensitive Matrices and Its Implications for Plastic Flow Localization and Crack-Tip Behavior," Ph.D. Thesis, The University of Michigan, Ann Arbor.

Jeong, H.-Y. and Pan, J., 1995, "A Macroscopic Constitutive Law for Porous Solids with Pressure-Sensitive Matrices and Its Implications to Plastic Flow Localization," *International Journal of Solids Structure*, Vol. 32, pp. 3669-3691.

Jeong, H.-Y. and Pan, J., 1996, "Crack-Tip Fields for Porous Solids with Pressure-Sensitive Matrices and for Rubber-Modified Epoxies," *Polymer Engineering and Science*, Vol. 36.

Lazzeri, A. and Bucknall, C. B., 1993, "Dilatational Bands in Rubber Toughened Polymers," *Journal of Materials Science*, Vol. 28, pp. 6799-6808.

Needleman, A. and Tvergaard, V., 1987, "An Analysis of Ductile Rupture Modes at a Crack Tip," *Journal of the Mechanics and Physics of Solids*, Vol. 35, pp. 151-183.

Pan, J., Saje, M. and Needleman, A., 1983, "Localization of Deformation in Rate Sensitive Porous Plastic Solids," *International Journal of Fracture*, Vol. 21, pp. 261-278.

Pearson, R. A. and Yee, A. F., 1986, "Toughening Mechanisms in Elastomer-Modified Epoxies, Part 2. Microscopic Studies," *Journal of Materials Science*, Vol. 21, pp. 2475-2488.

Pearson, R. A. and Yee, A. F., 1991, "Influence of Particle Size and Particle Size Distribution on Toughening Mechanisms in Rubber-Modified Epoxies," *Journal of Materials Science*, Vol. 26, pp. 3828-3844.

Rabinowitz, S., Ward, I. M. and Perry, J. S. C., 1970, "The Effect of Hydrostatic Pressure on the Shear Yield Behavior of Polymers," *Journal of Materials Science*, Vol. 5, pp. 29-39.

Sauer, J. A., Pae, K. D. and Bhateja, S. K., 1973, "Influence of Pressure on Yield and Fracture in Polymers," *Journal of Macromolecular Science – Physics*, Vol. B8, pp. 631-654.

Spitzig, W. A., Sober, R. J. and Richmond, O., 1975, "Pressure Dependence of Yielding and Associated Volume Expansion in Tempered Martensite," *Acta Metallurgica*, Vol. 19, pp. 1129-

1139.

Spitzig, W. A., Sober, R. J. and Richmond, O., 1976, "The Effect of Hydrostatic Pressure on the Deformation Behavior of Maraging and HY-80 Steels and Its Implications for Plasticity Theory," *Metallurgical Transactions*, Vol. 7A, pp. 1703-1710.

Spitzig, W. A. and Richmond, O., 1979, "Effect of Hydrostatic Pressure on the Deformation Behavior of Polyethylene and Polycarbonate in Tension and Compression," *Polymer Engineering and Science*, Vol. 19, pp. 1129-1139.

Sternstein, S. S. and Ongchin, L., 1969, "Yield Criteria for Plastic Deformation of Glassy High Polymers in General Stress Fields," *American Chemistry Society Polymer Preprint* 10, pp. 1117-1124.

Tandon, G. P. and Weng, G. J., 1988, "A Theory of Particle-Reinforced Plasticity," *Journal of the Applied Mechanics*, Vol. 55, pp. 126-135.

Tvergaard, V., 1981, "Influence of Voids on Shear Band Instabilities under Plane Strain Conditions," *International Journal of Fracture*, Vol. 17., pp. 389-407.

Tvergaard, V., 1982a, "On Localization in Ductile Materials Containing Spherical Voids," *International Journal of Fracture*, Vol. 18., pp. 237-252.

Tvergaard, V., 1982b, "Influence of Void Nucleation on Ductile Shear Fracture at a Free Surface," *Journal of the Mechanics and Physics of Solids*, Vol. 30, pp. 399-425.

Yee, A. F. and Pearson, R. A., 1986, "Toughening Mechanisms in Elastomer-Modified Epoxies, Part 1. Mechanical Studies," *Journal of Materials Science*, Vol. 21, pp. 2462-2474.

Yee, A. F., Li, D. and Li, X., 1993, "The Importance of Constraint Relief Caused by Rubber Cavitation in the Toughening of Epoxy," *Journal of Materials Science*, Vol. 28, pp. 6392-6398.

Yu, C.-S. and Shetty, D., 1989, "Transformation Zone Shape, Size, and Crack-Growth-Resistance [R-Curve] Behavior of Ceria-Partially-Stabilized Zirconia Polycrystals," *Journal of American Ceramic Society*, Vol. 72, pp. 921-928.

MD-Vol. 88, Polymeric Systems
ASME 1999

STRENGTHENING OF CRACKED REINFORCED CONCRETE BEAMS THROUGH CRACKFILLING WITH POLYMERIC MATERIALS

R. S. AYYAR and S. P. BURGUL

Sardar Patel College of Engineering
Mumbai, India

ABSTRACT
In this paper, an attempt is made to study the nature of weakening of R.C.C. beams due to crack formations and the extent of strengthening of the beam achieved through filling of the cracks by polymeric materials having high bond and flexural strength in tension. The   study of crack formation is carried out through a sequence of incremental loading till the stage of impending failure. The progressive   reduction in flexural rigidity with the growth of cracks is compared with the improvement in the same after crack filling.   A finite element method has been developed for this study

BEHAVIOUR OF REINFORCED CONCRETE BEAM UNDER INCREMENTAL LOADING
Plain   concrete beams are   unsuitable as flexural members because of the very low tensile strength of concrete in flexure. In a reinforced concrete beam, the tension caused by bending is mainly resisted by the steel reinforcement while the concrete resists   the compression,   provided relative slip is prevented between the two materials.

When the load on such a beam is gradually increased from zero to a magnitude that will cause the beam to fail, several stages of behaviour can be clearly distinguished. At low loads, as long as the maximum tensile stress in concrete is less than the modulus of rupture,   the entire concrete section is effective in resisting stresses both in tension and compression. In addition the reinforcement deforming the same amount as the adjacent concrete bonded to it is also subjected to tensile stress. At this stage all stresses in concrete are of small magnitude and are proportional to the corresponding strains.

As the load increases, the limiting tensile strength of concrete is reached and tensile cracks develop. These cracks propagate upward to or close to the level of neutral plane which, in turn, shifts upward with progressive cracking. The concrete in the cracked zone does not transmit any   tensile stresses and hence only the steel resists the entire tension. Upto moderate loads, stresses and strains continue   to be closely proportional but with further increase in load the stress – strain relation becomes non-linear.

Eventually the carrying capacity the beam is reached. Failure at this stage can be caused in one of two ways. With moderate amount of reinforcement, the steel yields   first stretching a large amount thereby causing the tensile cracks in concrete to widen and propagate upwards. Consequently the strains in the concrete in the compression zone increase to such a degree causing crushing of concrete and corresponding secondary compression failure. Effectively therefore attainment of yield point in steel determines the carrying capacity of the beam such yield failure is gradual and is preceded by visible signs of distress such as widening and lengthening of cracks and marked increase in deflection. On the other hand, if large amount of reinforcement or high strength steel is used, the compression strength of concrete may be exhausted before the steel starts yielding. Concrete fails by crushing which is sudden and occurs without any warning.

IDENTIFICATION OF THE PROBLEM AND METHOD OF ANALYSIS CHOSEN
As is obvious, the phenomenon of cracking is extremely complex and a reasonably good parameter to assess the structural behaviour of the beam at any stage of loading in its flexural rigidity. Flexural rigidity is the product of modulus of elasticity (E) and moment of Inertia (I). However, since concrete is a non-linear material, its E value is not constant for the entire stress – strain range. Again, after the cracks develop, the cross section of the beam changes from section to section with increasing loads and hence a constant section is not available along the   entire span to resist the load.   Such variation in resisting cross section cannot be accounted   for through a 'strength of materials' approach.

Also after crack filling the sectional properties of the beam change from section to section. For the crackfilled beam, there are three different materials viz. Concrete in the uncracked zone, polymeric material in the crackfilled zone and steel which may be embedded in concrete or in the polymeric material depending   on the location of the cracked zone. Determination of the flexural rigidity of such a beam is very complex and hence can be determined only if the problem is solved by the finite element technique.

With the finite element method, it is possible to change the material property wherever required and also know the variation of deflections, stresses and strains under the action of increasing loads at all the nodes of the interconnected elements. It is also possible to know where the beam will first crack and up to what depth under a given load by knowing the stress variation along the span. Hence, such a method is chosen for the present analysis.

ASSUMPTIONS MADE FOR THE ANALYSIS

1. There is perfect bond between steel and concrete.
2. There is perfect bond between steel, concrete and the crackfilling material.
3. The stress– strain behaviour of concrete in the compression zone is considered non-linear represented by a parabolic curve.
4. The modulus of elasticity (E) for concrete in the compression zone is taken as its secant modulus.
5. The stress-strain relation for concrete in the tension zone is assumed to be linear and the corresponding modulus of elasticity is taken to be constant given by $E = 5700\sqrt{fck}$, where f_{cr} is the modulus of rupture for concrete.
6. The modulus of elasticity for the crackfilling polymeric material is assumed to be constant.
7. The cracking pattern is idealised by corresponding whole number of cracked elements.
8. Steel is transformed into equivalent area of the material in which it is embedded (viz. Concrete or polymeric material)

IDEALIZATION OF REINFORCED CONCRETE BEAM

The finite element idealisation of a simply supported singly reinforced rectangular concrete beam is shown in Fig.1.

The concrete and steel reinforcement are both discretised into a system of rectangular elements. The beam is discretised into 1140 elements with a coarse grid near the support and refined grid near the centre of the span. The cross-section of the beam is shown in Fig.2. The tensile reinforcement is transformed into equivalent concrete area using modular ratio.

The analysis is carried out in the following sequential order.

1. Discretisation of a beam into rectangular elements and numbering of the same.
2. Numbering of nodal joints and calculation of the nodal co-ordinates.
3. Assembly of load matrix for the given load condition.
4. Assembly of global stiffness matrix.
5. Solution of linear equations which gives the unknown nodal displacement.
6. Calculation of nodal stresses and strains.

The only difference for all the three cases of uncracked, cracked and crackfilled beam is in the assembly of global stiffness matrix. The detail procedure used for such assembly in the three cases is as follows:

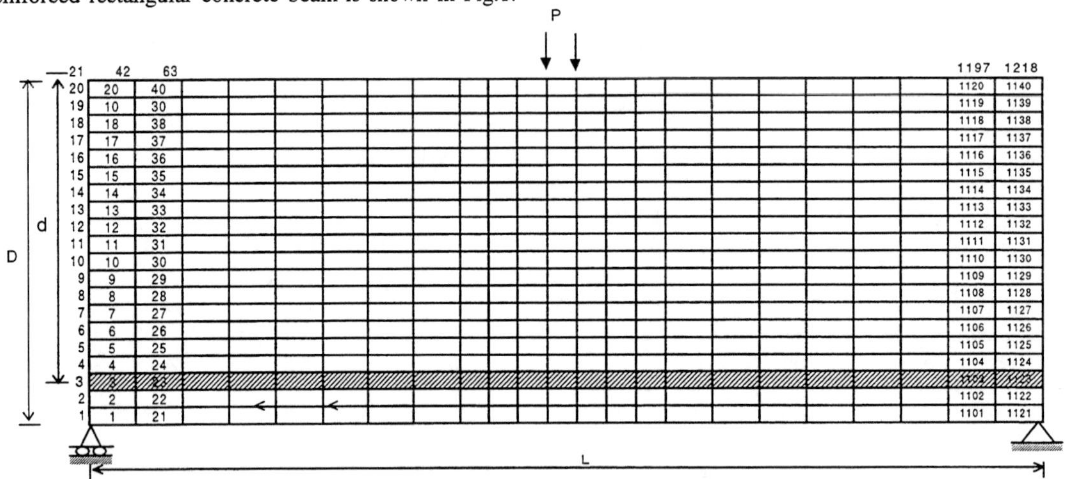

Fig. 1 Finite Element Idealization of a Singly Reinforced Concrete Beam

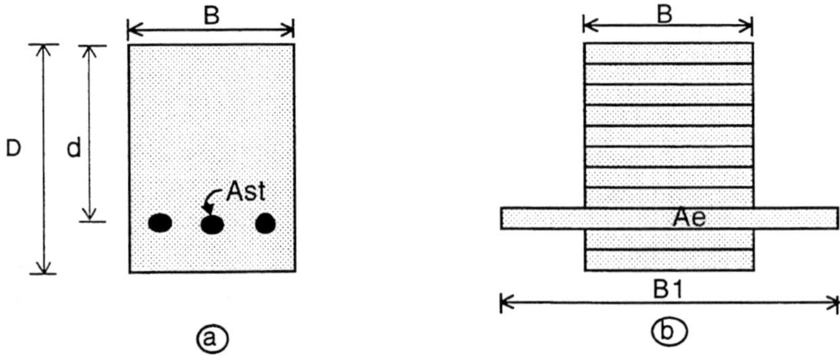

Fig. 2 Cross Section of the Beam
a. Actual section
b. Section for Finite Element Analysis

1. Assembly of stiffness matrix for uncracked beam :

The assembly of stiffness matrix for uncracked beam is simple compared to other two cases. The steel area is transformed into equivalent concrete area using modular ratio and that area is distributed along the width of the beam at the level of the reinforcement. The transformed R.C.C. beam section into equivalent concrete beam section is as shown in Fig 2(b). The equivalent concrete area for steel is

$$\text{'Ae'} = (m-1)Ast \qquad \ldots\ldots\ldots(1)$$

where,

$m = Es/Ec =$ modular ratio

$Es =$ Modulus of Elasticity of steel.

$Ec =$ Modulus of Elasticity of concrete

$Ast =$ Area of tensile steel.

Since the beam is divided into 20 layers along the depth, the transformed area of steel is distributed in one of those layer at the appropriate level. The increase in width of the beam (B1) at the level of steel is -

$$B1 = B + \frac{Ae.\,N}{D} \qquad \ldots\ldots\ldots\ldots(2)$$

Where,

$B =$ Width of the beam

$D =$ Depth of the beam

$N =$ No. of layer into which the beam depth is divided (which in the given case N=20).

While finding element stiffness matrix, the width of the beam is taken as 'B' for all the elements except for the elements at the level of the reinforcement for which it is taken as 'B1'. The assembly of the global stiffness matrix for the entire beam is carried out in half band matrix form as :

$$[K]\{\delta\} = \{F\} \qquad \ldots\ldots\ldots\ldots(3)$$

Where $\{\delta\}$ is unknown displacement vector. The simultaneous equations are solved by Gauss elimination method and unknown displacements are calculated. The strains at the nodes of each element are then obtained from the displacement vector $\{\delta\}$ using the strain matrix [B]. The stresses are then obtained from the strains using the corresponding stress-strain relationship.

The bending stresses calculated at the nodes of the elements give an overall view of stress distribution pattern along the depth and span of the beam. From this stress distribution pattern, it is easy to identify where the beam will crack and upto what depth for the given load condition by comparing stresses with the flexural strength of concrete in tension. The elements in which the stresses exceed the flexural strength of concrete in tension are considered to be cracked and stored in an array for subsequent study of cracked and crackfilled beam.

2. Assembly of stiffness matrix for cracked beam :

The elements in which the bending stresses in tension exceed the flexural stresses in tension exceed the flexural strength of concrete are assumed to have zero stiffness while assembling the stiffness matrix for the cracked beam. The cracked beam for finite element analysis is shown in Fig.3.

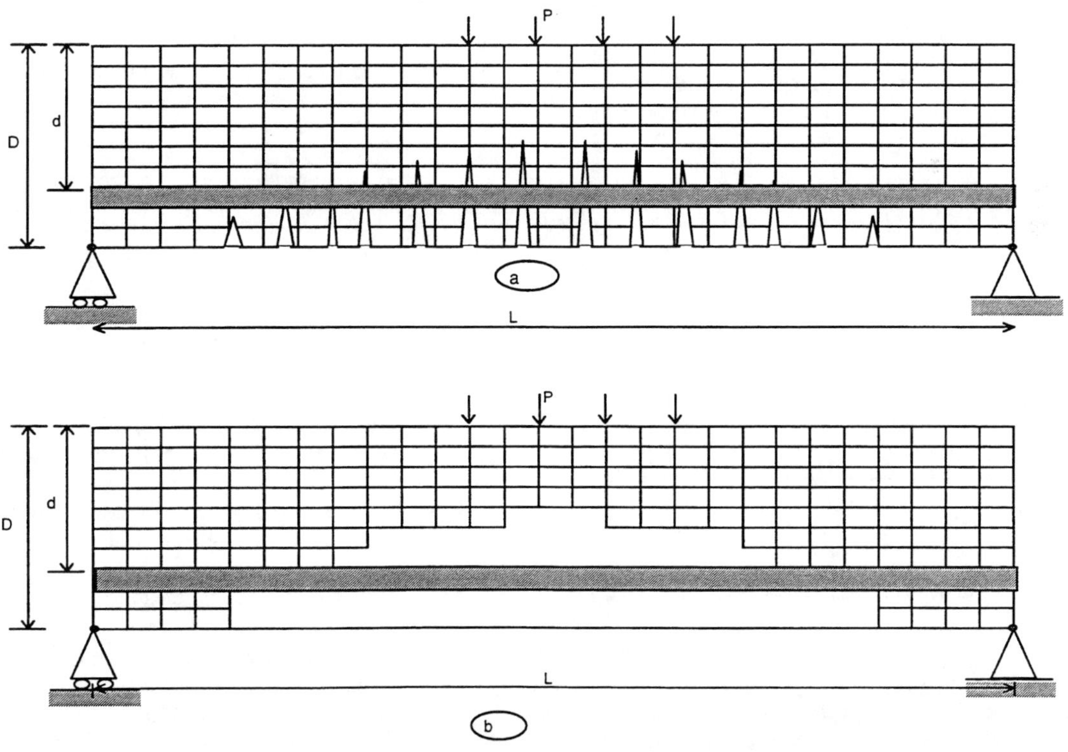

Fig . 3   a. Cracked Beam
              b. Finite Element Representation of Cracked Beam

The transforming of the R.C.C. section to equivalent concrete section is different for the cracked and uncracked sections. At uncracked section, the transformed beam section is the same as that for uncracked beam mentioned in case (1). i.e.

$$A_{e1} = (m-1) A_{st} \qquad \ldots\ldots\ldots\ldots(4)$$

At the cracked section, since it is assumed that there is no concrete, the equivalent area of concrete at the level of steel is given by

$$A_{e2} = m A_{st} \qquad \ldots\ldots\ldots\ldots(5)$$

Therefore the width of cross-section at the level of steel for uncracked zone is

$$B1 = B + \frac{A_{e1}N}{D} \qquad \ldots\ldots\ldots\ldots(6)$$

While for cracked zone it is

$$B1 = B + \frac{A_{e2} \cdot N}{D} \qquad \ldots\ldots\ldots\ldots(7)$$

By taking the correct width of the beam while finding the element stiffness matrix for each element, the assembly of the stiffness matrix for the entire beam is carried out and unknown displacements for the given load condition are calculated. Knowing the displacement vector, the corresponding strains at the nodal points are calculated using the strain-displacement matrix [B].

However, as the crack propagates with increase in loading, the stress-strain relation for concrete in the compression zone becomes non-linear as shown in Fig. 4. Hence, after getting the strain for each element in the compression zone, the corresponding stress is calculated from this curve (which is assumed to be parabolic). Using this stress and strain a modified value of E i.e. secant modulus of elasticity is calculated for that element which is used only for that element during the next incremental loading. However, for the uncracked concrete elements in the tension zone, a constant value of $E = 5700\sqrt{f_{ck}}$ is taken. In this way, the E value gets modified for each incremental loading until impending collapse stage is reached. The maximum permissible compressive strain in concrete is 0.002, which corresponds to the characteristic compressive strength of the material.

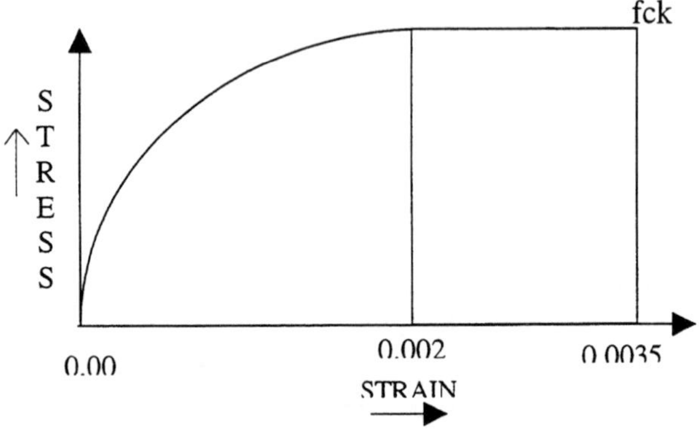

Fig.4 Stress-Strain Curve

1. Assembly of stiffness matrix for crackfilled beam :

For the crackfilled beam, there are three materials viz. concrete, steel and crackfilling material (which is high strength polymeric composite like epoxy and polyester concrete). So, the respective material properties must be considered while assembling the stiffness matrix. At the uncracked section, the transformed section is the same as that for case 1. At the

crackfilled section, there is new crackfilling material upto the limit of the crack and concrete above that. The reinforcement may be embedded either in concrete or in the crackfilling material depending on the depth of the crack at the corresponding section of the beam.

If steel lies embedded in concrete, then the equivalent concrete area is-

$$A_{e1} = (m-1)A_{st} \qquad \ldots\ldots\ldots\ldots(8)$$

But if steel lies embedded in the crackfilling material, the equivalent area of crackfilling material is

$$A_{e2} = (m_1-1)A_{st} \qquad \ldots\ldots\ldots\ldots(9)$$

where, $m_1 = Es/Ep$

m_1 = Modular ratio

Ep = Modulus of elasticity of the crackfilling material.

The transformed section for either case is shown in Fig.5

The assembly of the stiffness matrix is done for the entire beam and displacements are calculated by solving the simultaneous equations.

The vertical deflection of a beam is given by -

$$\delta = \frac{k}{EI} \qquad \ldots\ldots\ldots\ldots(10)$$

where, k = constant depending on the type of loading and support condition

EI = Flexural rigidity of the beam.

Hence, by calculating the vertical deflection with the help of the program for the uncracked, cracked and crackfilled beam, the corresponding variation in flexural rigidity can be determined.

ANALYSIS PROCEDURE

With the help of the program, the behaviour of the uncracked beam is first assessed both from the strength and stiffness points of view under the sequence of progressive incremental loading until the cracks develop. The cracks are then substituted by artificial notches of known width and depth and the loss of strength and stiffness determined in comparison to the uncracked beam. The process of incremental loading, the consequent growth of cracking zone and the corresponding structural behaviour of the beam are studied with the help of the program until a stage is reached just preceding impending collapse.

At this stage, the artificial notches substituted for the cracks are filled with polymeric composites having high strength properties in tension and bond and the strength and rigidity of the crackfilled beam are determined using the same program by changing the material properties in the crackfilled zone.

CASES OF SINGLY REINFORCED CONCRETE BEAM ANALYSED USING THE PROGRAM

The analysis of uncracked, cracked and crackfilled beam has been carried out for the following two load cases:-

1) Self weight + Central point load
2) Self-weight + Uniformly distributed load.

The beam analysed has the following properties.

i) Width of beam = 100mm
ii) Depth of beam = 200mm
iii) The beam is simply supported over a span of 2m.
iv) Grade of concrete = M20
v) Grade of steel = Fe415

For crackfilling the following polymeric composites considered.

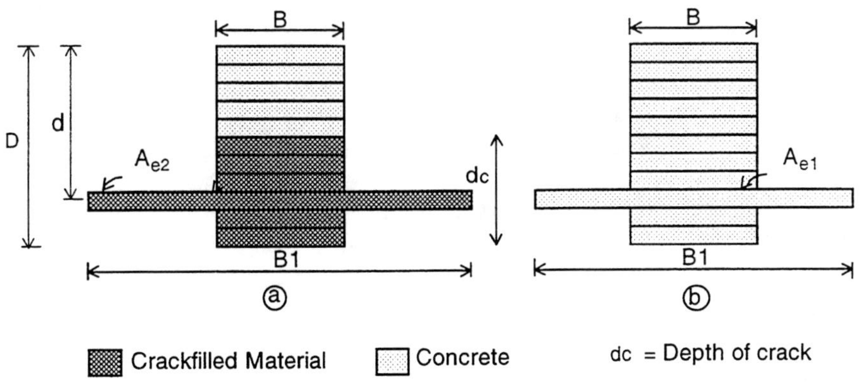

Fig.5 a. Crackfilled Section
b. Uncracked Section

Table:1; Mechanical Properties of Polymer Concrete (PC)

Sr. No.	Monomer	Polymer Dose, %of mass	Compressive Strength in MPa	Tensile Strength in MPa	Flexural Strength in MPa	Modulus of Elasticity, GPa
1	Concrete (M20)*	---	20	--	3.13	25.5
2	Polyester	1:10	117	12	37	32
3	Polyester	1:9	69	-	17	28
4	Epoxy+ Polyaminoamide	1:9	65	-	23	32
5	NMA-TMPTMA	1:15	137	10	22	35

*For comparison

The program starts with an initial load of 1kN (for central point load) and 1kN/m (for uniformly distributed load). For each incremental load, the deflections, strains and stresses are calculated in each element. The progressive variation in flexural rigidity is also computed by duly accounting for the propagation of cracks and the non-linear stress-strain relation for concrete in the compression zone.

OBSERVATIONS AND CONCLUSIONS

The variation of flexural rigidity with increase in loading for the two load cases considered is shown in Figs. 6 and 7. From these figs. it is noted that when the beam is in the uncracked stage the flexural rigidity remains almost constant. Once the cracks develop, there is a reduction in flexural rigidity. Subsequently with further increase in load, the flexural rigidity goes on decreasing until a stage is reached when the beam is no more able to sustain the applied load.

When the impending collapse stage is reached, the entire concrete in the cracked zone is removed and replaced by high strength polymeric concrete like epoxy and polyester concrete. The beam is loaded once again with incremental loading until the stage of impending collapse. But the behaviour of the crackfilled beam is observed to be different from that of the original reinforced concrete beam.

As the strength of polymer is very high compared to ordinary concrete (M20), the initial cracks in the crackfilled beam develop not in the crackfilled zone which is subjected to maximum bending moment but in the adjoining zones of initially uncracked concrete. These cracks propagate upward with increasing load and the beam fails due to the stress in compression zone exceeding the permissible limit for concrete. The load carrying capacity of the crackfilled beam is higher than that of the original R.C.C. beam. The flexural rigidity at failure of the crackfilled beam is also not the same as that of the original R.C.C. beam. In the R.C.C. beam, the failure is characterised by yielding of steel while in the crackfilled beam, it is caused by crushing of concrete in the compression zone. Such distinction in the failure patterns is the main reason for the difference in the flexural rigidities.

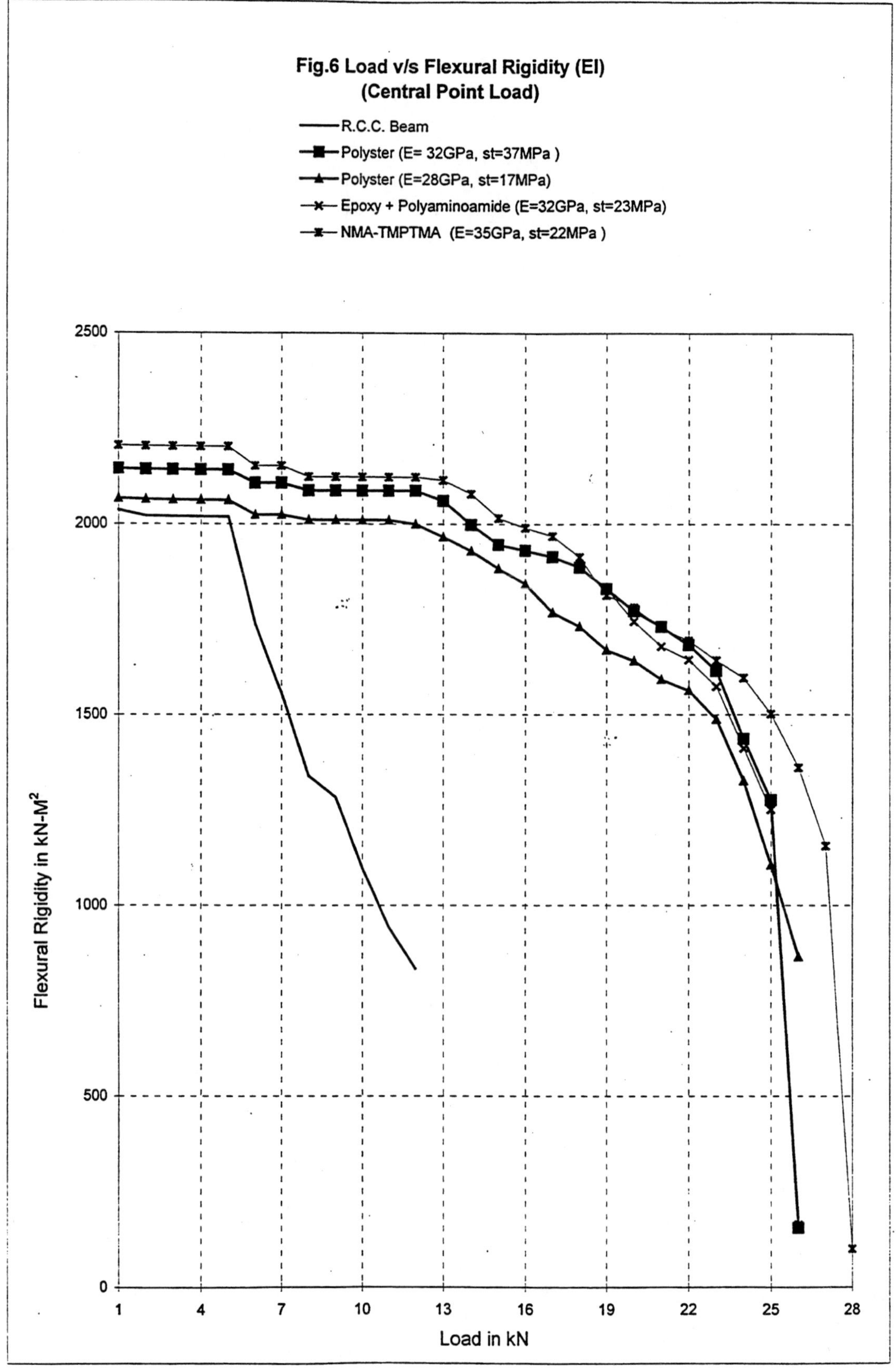

Fig.6 Load v/s Flexural Rigidity (EI)
(Central Point Load)
R.C.C. Beam
Polyster (E= 32GPa, st=37MPa)
Polyster (E=28GPa, st=17MPa)
Epoxy + Polyaminoamide (E=32GPa, st=23MPa)
NMA-TMPTMA (E=35GPa, st=22MPa)
2500
2000
1500
1000
500
0
Flexural Rigidity in kN-M^2
1
4
7
10
13
16
19
22
25
28
Load in kN

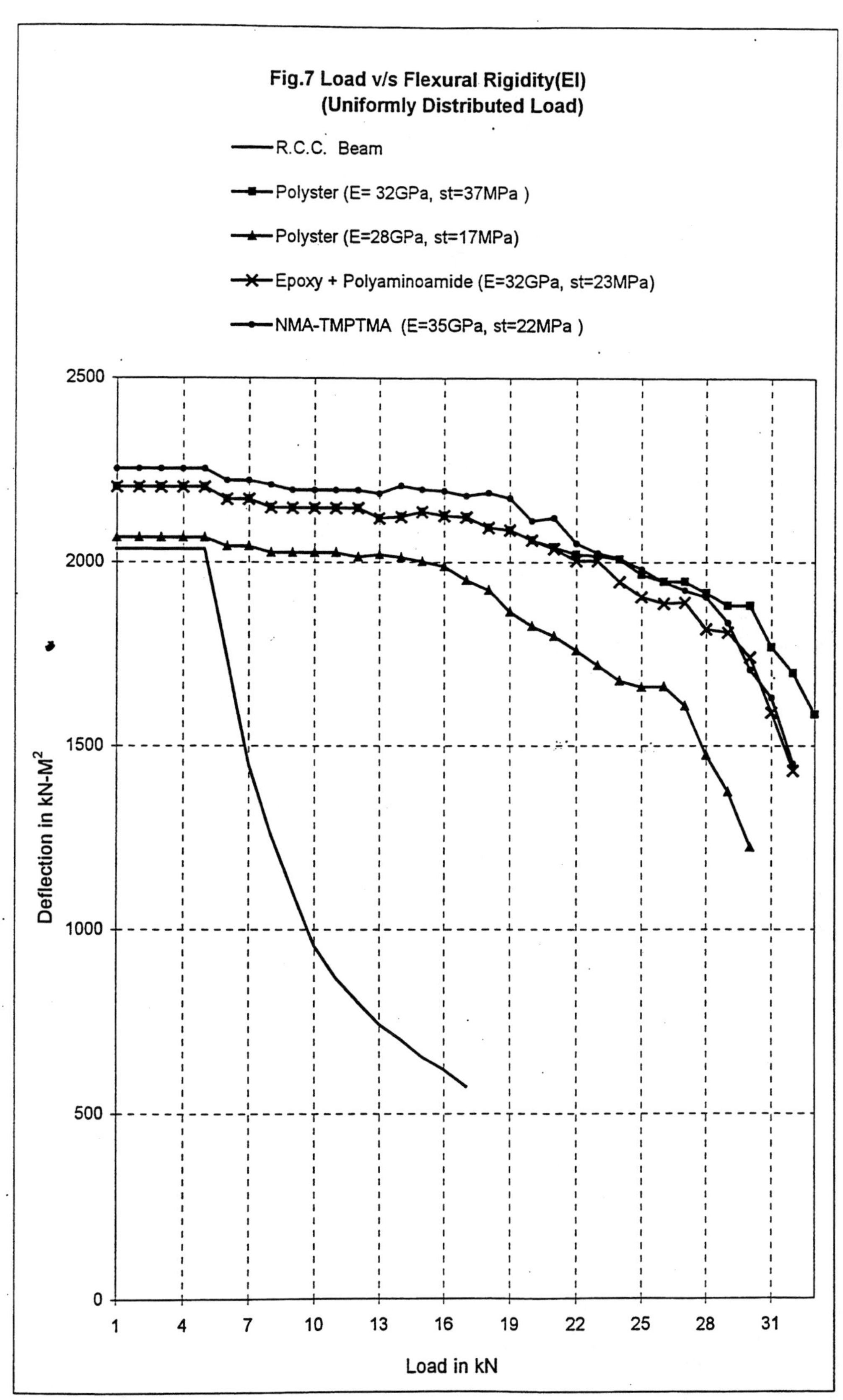

Fig.7 Load v/s Flexural Rigidity(EI)
(Uniformly Distributed Load)
R.C.C. Beam
Polyster (E= 32GPa, st=37MPa)
Polyster (E=28GPa, st=17MPa)
Epoxy + Polyaminoamide (E=32GPa, st=23MPa)
NMA-TMPTMA (E=35GPa, st=22MPa)
2500
2000
1500
1000
500
0
Deflection in kN-M^2
1
4
7
10
13
16
19
22
25
28
31
Load in kN

The load carrying capacity also shows a marked increase in the crackfilled beam compared to the original R.C.C. beam. Thus for a central point load the collapse load for the crackfilled beam is nearly 28kN (when the cracks are filled with NMA+TMPTMA) as against 12kN for the R.C.C. beam (Fig.6). for the case of uniformly distributed load, the corresponding value is 33kN (for crackfilling with polyester) as against 17kN for the R.C.C. beam (Fig.7)

GENERAL REMARKS BASED ON THE ANALYSIS

While cracks are unavoidable in the tension zone of R.C.C. beams the safety of the beam depends essentially on the nature and extent of such crack formation. When the cracks cross the permissible limits of safety, the best rehabilitation measure is the filling of these cracks by high strength materials like polymer composites. But such crack filling has to be followed by a reasonably correct assessment of the improvement in flexural rigidity of the crack filled beam and the corresponding increase in factor of safety. The finite element analysis is developed gives a reasonably reliable method for such assessment through modelling the uncracked, cracked and crack filled beams under any stage of loading by changing the material properties of the elements at the appropriate zones.

REFERENCES

1. ACI Committee 224: 'Control of Cracking in Concrete Structure' Journal ACI, vol.69, No.12 Dec. 1972. pp:717-753.

2. A.H. Nilson: 'Non-linear Analysis of Reinforced Concrete by the Finite Element Method', Journal A.C.I., Vol.65, No.9 Sept.1968. pp :758-765

3. IS 456-1978: Indian Standard Code of Practice for Plain and Reinforced Concrete (Third Revision)

4. S.K. Manjrekar: 'Polymers in Concrete: Mechanism, Properties and Applications', Indian Concrete Journal, March-1992,pp-127.

AUTHOR INDEX

MD-Vol. 88
Polymeric Systems